AMERICAN MATHEMATICAL SOCIETY
COLLOQUIUM PUBLICATIONS
VOLUME XIX

FOURIER TRANSFORMS
IN THE
COMPLEX DOMAIN

BY

The late RAYMOND E. A. C. PALEY

Late Fellow of Trinity College, Cambridge

AND

NORBERT WIENER

Professor of Mathematics at the Massachusetts Institute of Technology

PUBLISHED BY THE
AMERICAN MATHEMATICAL SOCIETY
1934

Seventh Printing 1973

International Standard Book Number 0-8218-1019-7

Library of Congress Catalog Card Number 35-3273

PHOTOLITHOPRINTED BY CUSHING - MALLOY, INC.
ANN ARBOR, MICHIGAN, UNITED STATES OF AMERICA

R. E. A. C. Paley, 1907–1933

Dedicated by the surviving author

to

PROFESSORS G. H. HARDY *and* J. E. LITTLEWOOD

the teachers of us both

Preface

The present book represents a definitive statement of the results obtained by the late Mr. R. E. A. C. Paley and myself during Mr. Paley's year as Rockefeller Fellow at the Massachusetts Institute of Technology (1932–1933). Mr. Paley was killed on April 7 in a skiing accident in the Canadian Rockies, during a short vacation which he had taken from our joint work. I have written elsewhere of the great loss to mathematics by his death; here let me only state the condition in which our joint work was left. Our method of collaboration had been most informal. We had worked together with a blackboard before us, and when we had covered it with our joint comments, one or the other would copy down what was relevant, and reduce it to a preliminary written form. Most of our work went through several versions, in writing which both authors took part. Even in that part of the research committed to writing since Mr. Paley's death, it is completely impossible to determine how much is new and how much is a reminiscence of our many conversations.

A part of our work was published in the form of a series of notes in the Transactions of the American Mathematical Society. This work covered a great variety of topics, but was unified by the central idea of the application of the Fourier transform in the complex domain. I had long been convinced of the importance of the Fourier-Mellin transforms as a tool in analysis. Their introduction is of course no novelty, but I know of no systematic development of their methodical use. Perhaps the nearest approach to such a development is to be found in the researches of H. Bohr, Jessen and Besicovitch on almost periodic functions in the complex domain. However, nobody seems to have realized anything like the scope of the method. With its aid, we were able to attack such diverse analytic questions as those of quasi-analytic functions, of Mercer's theorem on summability, of Milne's integral equation of radiative equilibrium, of the theorems of Münz and Szász concerning the closure of sets of powers of an argument, of Titchmarsh's theory of entire functions of semi-exponential type with real negative zeros, of trigonometric interpolation and developments in polynomials of the form

$$\sum_1^N A_n e^{i\lambda_n x},$$

of lacunary series, of generalized harmonic analysis in the complex domain, of the zeros of random functions, and many others. We came to believe that an analytic method of such scope is entitled to an independent treatise.

The American Mathematical Society has done me the honor of requesting me to deliver the Williamstown Colloquium Lectures for 1934. While such lectures have not previously been an account of collaborative work, my best available work has been collaborative, and I have offered it for the lectures in question.

I wish to thank the American Mathematical Society for its invitation, and for its acceptance of our plans. I wish to thank my students, Messrs S. S. Saslaw, H. Malin, and N. Levinson, for most valuable and painstaking work of revision, compilation, and criticism. Mr. Levinson, in particular, has added much to the content of Chapter I. I wish to thank my colleague, Professor Eberhard Hopf, for permission to incorporate into this book the material of §17, which was our joint work. Furthermore, in my own name and in the name of my dead co-author, I wish to thank Professor J. D. Tamarkin of Brown University for his untiring encouragement, advice, and criticism, without which this book would not have come into existence.

NORBERT WIENER.

MASSACHUSETTS INSTITUTE OF TECHNOLOGY,
CAMBRIDGE, MASSACHUSETTS, MARCH 1, 1934.

Contents

Introduction

1. Plancherel's Theorem. In a book such as the present, unified rather by the repeated use of a number of methods than by a great homogeneity of content, it is quite necessary to start with a brief account of the background of scientific knowledge presupposed and a tabulation of the principal methods. The background of knowledge presupposed in the greater part of this treatise is roughly that covered in Titchmarsh's very useful *Theory of Functions*. The tools most used are the following:

(1) Integration by parts, and other similar inversions of the order of an absolutely convergent double integral;

(2) The "mutilation" of the function: that is, the replacement of a function by a function identical with the first over a finite range, and vanishing outside that range;

(3) The Schwarz inequality

$$\left[\int_a^b | f(x) g(x) | dx\right]^2 \leqq \int_a^b | f(x) |^2 dx \int_a^b | g(x) |^2 dx, \tag{1.01}$$

and similar inequalities for sums, series, etc.;

(4) The Weyl form of the Riesz-Fischer theorem, to the effect that if a sequence of functions $\{f_n(x)\}$ of L_2 converges in the mean in the sense

$$\lim_{m,n\to\infty} \int_a^b | f_m(x) - f_n(x) |^2 dx = 0, \tag{1.02}$$

then there exists a function $f(x)$ of L_2 to which the sequence converges in the mean in the sense

$$\lim_{m\to\infty} \int_a^b | f_m(x) - f(x) |^2 dx = 0. \tag{1.03}$$

(5) The theorem that if a sequence of functions converges in the mean to one limit, and converges in the ordinary sense to another, then these two limits differ at most over a null set;

(6) Methods of summability and averaging, in particular theorems of Abelian and Tauberian types:*

(7) Methods depending on the Plancherel and the Parseval theorems concerning Fourier transforms.†

* Cf. Wiener, *The Fourier Integral and Certain of its Applications*, Cambridge, 1933. The Tauberian theorems of this book are not to be found in the book of Titchmarsh.

† A good elementary study of such questions is to be found in S. Bochner's *Vorlesungen über Fouriersche Integrale*, Leipzig, 1932. The treatment in Wiener's book (see above) is slightly more advanced.

Throughout this book we shall assume on the part of the reader familiarity with the theory and use of the Lebesgue integral and with the appropriate notations. In particular, we shall make repeated use of the notation L to represent the class of measurable, absolutely integrable functions, and of the notation L_p for the class of those measurable functions, the pth power of whose modulus is integrable. However, we shall have little to do with other classes than L and L_2.

In the theory of the Fourier integral, the fundamental theorem for the class L_2 is that of Plancherel. It reads as follows:

PLANCHEREL'S THEOREM. *Let $f(x)$ belong to L_2 over $(-\infty, \infty)$. Then there exists a function $g(u)$ belonging to L_2 over $(-\infty, \infty)$, and such that*

$$\lim_{A\to\infty} \int_{-\infty}^{\infty} \left| g(u) - (2\pi)^{-1/2} \int_{-A}^{A} f(x)\, e^{iux}\, dx \right|^2 du = 0. \tag{1.04}$$

Furthermore

$$\int_{-\infty}^{\infty} |\, g(u)\,|^2\, du = \int_{-\infty}^{\infty} |\, f(x)\,|^2\, dx, \tag{1.05}$$

and

$$\lim_{A\to\infty} \int_{-\infty}^{\infty} \left| f(x) - (2\pi)^{-1/2} \int_{-A}^{A} g(u)\, e^{-iux}\, du \right|^2 dx = 0. \tag{1.06}$$

The function $g(u)$ is called the Fourier Transform of $f(x)$. It is determined except over a set of points of zero measure.

In case

$$h(u) = (2\pi)^{-1/2} \int_{-\infty}^{\infty} f(x)\, e^{iux}\, dx \tag{1.07}$$

exists, we have $g(u) = h(u)$ almost everywhere.

An important corollary of Plancherel's theorem is

PARSEVAL'S THEOREM. *Let $f_1(x)$ and $f_2(x)$ both belong to L_2, and let them have the Fourier transforms $g_1(u)$ and $g_2(u)$, respectively. Then*

$$\int_{-\infty}^{\infty} g_1(u)\, g_2(u)\, du = \int_{-\infty}^{\infty} f_1(x)\, f_2(-x)\, dx. \tag{1.08}$$

In particular,

$$\int_{-\infty}^{\infty} g_1(u)\, g_2(u)\, e^{-iux}\, du = \int_{-\infty}^{\infty} f_1(y)\, f_2(x-y)\, dy. \tag{1.09}$$

Thus, if $g_1(u)\, g_2(u)$ belongs to L_2 as well as both its factors, it is the Fourier transform of

$$(2\pi)^{-1/2} \int_{-\infty}^{\infty} f_1(y)\, f_2(x-y)\, dy. \tag{1.10}$$

This will also be true whenever $f_1(x)$, $f_2(x)$, and (1.10) all belong to L_2.

2. **The Fourier transform of a function vanishing exponentially.** Let us suppose that $f(x)$ is measurable, and of summable square over any finite interval. Let

$$f(x) = \begin{cases} O(e^{-\mu x}) & [x \to \infty]; \\ O(e^{\lambda x}) & [x \to -\infty]. \end{cases} \tag{2.1}$$

In case $-\lambda < \sigma < \mu$, the Fourier transform of $f(x)\, e^{\sigma x}$ will then be

$$F(\sigma, t) = (2\pi)^{-1/2} \int_{-\infty}^{\infty} f(x)\, e^{(\sigma+it)x}\, dx. \tag{2.2}$$

However, this converges absolutely and uniformly over every range $-\lambda + \epsilon < \sigma < \mu - \epsilon$. Thus by a well known theorem from the theory of functions of a complex variable,

$$F(\sigma + it) = F(\sigma, t) \tag{2.3}$$

will be an analytic function of $\sigma + it$ over the interior of the strip $-\lambda < \sigma < \mu$. Furthermore, over any strip $-\lambda + \epsilon < \sigma < \mu - \epsilon$ we shall have

$$\begin{aligned} \int_{-\infty}^{\infty} | F(\sigma, t) |^2\, dt &= \int_{-\infty}^{\infty} | f(x) |^2\, e^{2\sigma x}\, dx \\ &< \text{const.} \int_0^{\infty} e^{-2\epsilon x}\, dx + \text{const.} \int_{-\infty}^{0} e^{2\epsilon x}\, dx = \text{const.} \end{aligned} \tag{2.4}$$

We have thus proved

THEOREM I. *If $f(x)$ is measurable, of summable square over every finite interval, and a function satisfying* (2.1), *for $-\lambda < \mu$, then* (2.2) *defines a function $F(\sigma + it)$ analytic over the interior of the strip $-\lambda < \sigma < \mu$; and over any interior strip $-\lambda + \epsilon \leqq \sigma \leqq \mu - \epsilon$, $\int_{-\infty}^{\infty} | F(\sigma + it) |^2\, dt$ is bounded.*

3. **The Fourier transform of a function in a strip.** Let $F(\sigma + it)$ be a function of the complex variable $s = \sigma + it$, which is analytic in and on the boundary of the strip $-\lambda \leqq \sigma \leqq \mu$, and let

$$\int_{-\infty}^{\infty} | F(\sigma + it) |^2\, dt < \text{const.} \qquad [-\lambda \leqq \sigma \leqq \mu]. \tag{3.01}$$

Then by Cauchy's theorem, if A is large enough and $-\lambda < \sigma < \mu$,

$$F(s) = \frac{1}{2\pi i} \left[\int_{-\lambda+Ai}^{-\lambda-Ai} + \int_{-\lambda-Ai}^{\mu-Ai} + \int_{\mu-Ai}^{\mu+Ai} + \int_{\mu+Ai}^{-\lambda+Ai} \right] \frac{F(z)}{z - s}\, dz. \tag{3.02}$$

By a further integration, if B is large enough,

$$F(s) = \frac{1}{2\pi i} \int_B^{B+1} dA \left[\int_{-\lambda+Ai}^{-\lambda-Ai} + \int_{-\lambda-Ai}^{\mu-Ai} + \int_{\mu-Ai}^{\mu+Ai} + \int_{\mu+Ai}^{-\lambda+Ai} \right] \frac{F(z)}{z - s}\, dz. \tag{3.03}$$

Now,

$$\left|\frac{1}{2\pi i}\int_B^{B+1} dA \int_{\mu+Ai}^{-\lambda+Ai} \frac{F(z)}{z-s}\,dz\right| = \frac{1}{2\pi}\left|\int_{-\lambda}^{\mu} dz \int_B^{B+1} \frac{F(z+Ai)}{z+Ai-s}\,dA\right|$$

$$(3.04)\qquad \leqq \frac{1}{2\pi}\int_{-\lambda}^{\mu} dz \left\{\int_B^{B+1} |F(z+Ai)|^2\,dA \int_B^{B+1} \frac{dA}{|z+Ai-s|^2}\right\}^{1/2}$$

$$\leqq \text{const.} \int_{-\lambda}^{\mu} d\sigma \left\{\int_B^{B+1} |F(\sigma+it)|^2\,dt\right\}^{1/2}.$$

By (3.01) it follows that

$$(3.05)\qquad \begin{cases} \left\{\int_B^{B+1} |F(\sigma+it)|^2\,dt\right\}^{1/2} < \text{const.}, \\ \lim_{B\to\infty} \left\{\int_B^{B+1} |F(\sigma+it)|^2\,dt\right\}^{1/2} = 0. \end{cases}$$

It is a familiar theorem in the theory of the Lebesgue integral, that if a sequence of integrable functions converges boundedly to a limit, and the integral of that limit exists, then the limit of the integral of a function of the sequence is the integral of the limit. Thus

$$(3.06)\qquad \lim_{B\to\infty} \frac{1}{2\pi i}\int_B^{B+1} dA \int_{\mu+Ai}^{-\lambda+Ai} \frac{F(z)}{z-s}\,dz = 0.$$

Similarly,

$$(3.07)\qquad \lim_{B\to\infty} \frac{1}{2\pi i}\int_B^{B+1} dA \int_{-\lambda-Ai}^{\mu-Ai} \frac{F(z)}{z-s}\,dz = 0.$$

Thus

$$(3.08)\qquad \begin{aligned} F(s) &= \lim_{B\to\infty} \frac{1}{2\pi}\int_B^{B+1} dA \int_{-A}^{A} \left\{\frac{F(\mu+iy)}{\mu+iy-s} - \frac{F(-\lambda+iy)}{-\lambda+iy-s}\right\} dy \\ &= \lim_{B\to\infty} \left\{\frac{1}{2\pi}\int_{-B}^{B} \frac{F(\mu+iy)}{\mu+iy-s}\,dy - \frac{1}{2\pi}\int_{-B}^{B} \frac{F(-\lambda+iy)}{-\lambda+iy-s}\,dy \right. \\ &\quad + \frac{1}{2\pi}\int_B^{B+1} \frac{(B+1-y)}{\mu+iy-s} F(\mu+iy)\,dy \\ &\quad - \frac{1}{2\pi}\int_B^{B+1} \frac{(B+1-y)\,F(-\lambda+iy)}{-\lambda+iy-s}\,dy \\ &\quad + \frac{1}{2\pi}\int_{-B-1}^{-B} \frac{(B+1+y)\,F(\mu+iy)}{\mu+iy-s}\,dy \\ &\quad \left. - \frac{1}{2\pi}\int_{-B-1}^{-B} \frac{(B+1+y)\,F(-\lambda+iy)}{-\lambda+iy-s}\,dy \right\}. \end{aligned}$$

Now,

$$\left|\frac{1}{2\pi}\int_B^{B+1}\frac{(B+1-y)\,F(\mu+iy)}{\mu+iy-s}\,dy\right| \leqq \frac{1}{2\pi}\left\{\int_B^{B+1}\left|\frac{B+1-y}{\mu+iy-s}\right|^2 dy\right.$$

$$\tag{3.09} \left.\times \int_B^{B+1} |F(\mu+iy)|^2\,dy\right\}^{1/2}$$

$$\leqq \text{const.}\left\{\int_B^{B+1} |F(\mu+iy)|^2\,dy\right\}^{1/2}$$

It follows at once from (3.01) that

$$\tag{3.10} \lim_{B\to\infty}\frac{1}{2\pi}\int_B^{B+1}\frac{(B+1-y)\,F(\mu+iy)}{(\mu+iy-s)}\,dy = 0.$$

A similar argument will enable us to eliminate three more terms from (3.08), and we get

$$\tag{3.11} F(s) = \frac{1}{2\pi}\int_{-\infty}^{\infty}\frac{F(\mu+iy)}{\mu+iy-s}\,dy - \frac{1}{2\pi}\int_{-\infty}^{\infty}\frac{F(-\lambda+iy)}{-\lambda+iy-s}\,dy.$$

This is a sufficiently important result to be numbered as a theorem. We have

THEOREM II. *Let $F(s)$ be analytic over $-\lambda \leqq \sigma \leqq \mu$, and let* (3.01) *hold over this region. Then if s is interior to this region,* (3.11) *is valid.*

The use of the Schwarz inequality gives us

$$\tag{3.12} \left|\frac{1}{2\pi}\int_{-\infty}^{\infty}\frac{F(\mu+iy)}{(\mu+iy-s)}\,dy\right| \leqq \frac{1}{2\pi}\left\{\int_{-\infty}^{\infty}|F(\mu+iy)|^2\,dy\int_{-\infty}^{\infty}\frac{dy}{|\mu+iy-s|^2}\right\}^{1/2},$$

or

THEOREM III. *Under the hypothesis of Theorem* II, *$F(s)$ is bounded over any region $-\lambda+\epsilon \leqq \sigma \leqq \mu-\epsilon$.*

Let us put

$$\tag{3.13} f(\sigma,x) = (2\pi)^{-1/2}\,\text{l.i.m.}\int_{-A}^{A} F(\sigma+it)\,e^{-itx}\,dt.$$

Let us also put

$$\tag{3.14} \phi(x) = 0\ [x<0]; \qquad \phi(x) = e^{-\alpha x}\ [x>0]; \qquad \alpha > 0.$$

We shall have

$$\tag{3.15} \int_{-\infty}^{\infty}\phi(x)\,e^{ixy}\,dx = \int_0^{\infty} e^{ixy-\alpha x}\,dx = \frac{1}{\alpha-iy}.$$

Plancherel's theorem yields

$$\tag{3.16} \underset{A\to\infty}{\text{l.i.m.}}\,\frac{1}{2\pi}\int_{-A}^{A}\frac{e^{-ixy}}{\alpha-iy}\,dy = \begin{cases}0 & [x<0]\\ e^{-\alpha x} & [x>0]\end{cases} \text{ if } \alpha>0.$$

Similarly, if $\alpha < 0$,

$$\operatorname*{l.i.m.}_{A\to\infty} \frac{1}{2\pi} \int_{-A}^{A} \frac{e^{ixy}}{\alpha - iy}\, dy = \begin{cases} -e^{-\alpha x} & [x < 0] \\ 0 & [x > 0]. \end{cases} \tag{3.17}$$

Thus by the Parseval theorem, since $-\lambda + \epsilon < \sigma < \mu - \epsilon$,

$$\int_{-\infty}^{0} f(-\lambda, x)\, e^{(\sigma+\lambda)x}\, e^{itx}\, dx = \frac{1}{(2\pi)^{1/2}} \int_{-\infty}^{\infty} \frac{F(-\lambda + iy)}{[-\lambda - \sigma - (t - y)i]}\, dy. \tag{3.18}$$

In exactly the same way,

$$\int_{0}^{\infty} f(\mu, x)\, e^{(\sigma-\mu)x}\, e^{itx}\, dx = \frac{1}{(2\pi)^{1/2}} \int_{-\infty}^{\infty} \frac{F(\mu + iy)}{[\mu - \sigma - (t - y)i]}\, dy. \tag{3.19}$$

Thus by (3.11)

$$F(s) = \frac{1}{(2\pi)^{1/2}} \int_{-\infty}^{0} f(-\lambda, x)\, e^{(\sigma+\lambda)x}\, e^{itx}\, dx + \frac{1}{(2\pi)^{1/2}} \int_{0}^{\infty} f(\mu, x)\, e^{(\sigma-\mu)x}\, e^{itx}\, dx. \tag{3.20}$$

Another use of Plancherel's theorem gives us

$$f(\sigma, x) = \begin{cases} f(-\lambda, x)\, e^{(\sigma+\lambda)x} & [-\infty < x < 0]; \\ f(\mu, x)\, e^{(\sigma-\mu)x} & [0 < x < \infty]; \end{cases} \tag{3.21}$$

and consequently if for a particular α

$$f(x) = f(\alpha, x) e^{-\alpha x} \qquad [-\lambda < \alpha < \mu], \tag{3.22}$$

we get

$$f(\sigma, x) = f(x) e^{\sigma x} \qquad [-\lambda < \sigma < \mu]. \tag{3.23}$$

An immediate result is that

$$\lim_{\sigma, \sigma_1 \to \mu - 0} \int_{A}^{B} | f(\sigma, x) - f(\sigma_1, x) |^2\, dx = 0. \tag{3.24}$$

Moreover, if B is positive and A negative,

$$\int_{B}^{\infty} | f(\sigma, x) |^2\, dx \leqq \int_{B}^{\infty} | f(\mu, x) |^2\, dx, \tag{3.25}$$

and

$$\int_{-\infty}^{A} | f(\sigma, x) |^2\, dx \leqq \int_{-\infty}^{A} | f(-\lambda, x) |^2\, dx. \tag{3.26}$$

As a consequence,

$$\lim_{\sigma, \sigma_1 \to \mu - 0} \int_{-\infty}^{\infty} | f(\sigma, x) - f(\sigma_1, x) |^2\, dx = 0. \tag{3.27}$$

Thus by Weyl's lemma to the Riesz-Fischer theorem, there exists in L_2 a function $f_1(x)$ such that

$$\lim_{\sigma \to \mu - 0} \int_{-\infty}^{\infty} | f(\sigma, x) - f_1(x) |^2 \, dx = 0. \tag{3.28}$$

By a Fourier transformation and the use of Plancherel's theorem, there exists a function $F_1(t)$ such that

$$\lim_{\sigma \to \mu - 0} \int_{-\infty}^{\infty} | F(\sigma + it) - F_1(t) |^2 \, dt = 0. \tag{3.29}$$

However, we already know that

$$\lim_{\sigma \to \mu - 0} F(\sigma + it) = F(\mu + it), \tag{3.30}$$

and hence,

$$\lim_{\sigma \to \mu - 0} \int_{-\infty}^{\infty} | F(\sigma + it) - F(\mu + it) |^2 \, dt = 0. \tag{3.31}$$

By a further Fourier transformation,

$$\lim_{\sigma \to \mu - 0} \int_{-\infty}^{\infty} | f(\sigma, x) - f(\mu, x) |^2 \, dx = 0. \tag{3.32}$$

Similarly,

$$\lim_{\sigma \to -\lambda + 0} \int_{-\infty}^{\infty} | f(\sigma, x) - f(-\lambda, x) |^2 \, dx = 0. \tag{3.33}$$

However,

$$\lim_{\sigma \to \mu} f(\sigma, x) = f(-\lambda, x)\, e^{(\mu+\lambda)x} \qquad [-\infty < x < 0], \tag{3.34}$$

and

$$\lim_{\sigma \to -\lambda} f(\sigma, x) = f(\mu, x)\, e^{-(\mu+\lambda)x} \qquad [0 < x < \infty]. \tag{3.35}$$

Thus with the possible exception of a null set,

$$f(\sigma, x) = f(\mu, x)e^{(\sigma-\mu)x} = f(-\lambda, x)e^{(\sigma+\lambda)x} \qquad [-\infty < x < \infty]. \tag{3.36}$$

This yields us

THEOREM IV. *Under the hypotheses of Theorem* II, *there exists a measurable function* $f(x)$, *such that*

$$\int_{-\infty}^{\infty} | f(x) |^2 \, e^{2\mu x} \, dx < \infty, \qquad \int_{-\infty}^{\infty} | f(x) |^2 \, e^{-2\lambda x} \, dx < \infty, \tag{3.37}$$

and that over the closed interval $-\lambda \leqq \sigma \leqq \mu$,

$$F(\sigma + it) = \underset{A \to \infty}{\text{l.i.m.}}\, (2\pi)^{-1/2} \int_{-A}^{A} f(x)\, e^{x(\sigma+it)} \, dx. \tag{3.38}$$

It follows from Theorems I and IV that the extreme boundaries of the interval over which $F(\sigma + it)$ belongs uniformly to L_2 as a function of t are given by the boundaries of convergence of the integral

$$\int_{-\infty}^{\infty} | f(x) |^2 e^{2\sigma x} \, dx.$$

4. The Fourier transform of a function in a half-plane. In particular, $F(\sigma + it)$ will belong to L_2 in every ordinate of a right half-plane when and only when

$$\int_{-\infty}^{\infty} | f(x) |^2 e^{2\sigma x} \, dx < \infty \tag{4.01}$$

for all sufficiently large σ, and will belong to L_2 *uniformly* in such a half-plane when and only when

$$\lim_{\sigma \to \infty} \int_{-\infty}^{\infty} | f(x) |^2 e^{2\sigma x} \, dx < \infty. \tag{4.02}$$

This latter contingency can only occur when $f(x)$ vanishes almost everywhere for positive values of its argument. Otherwise there will be some interval (a, b) $[b > a > 0]$ over which

$$\int_{a}^{b} | f(x) |^2 \, dx = I > 0. \tag{4.03}$$

We shall then have

$$\int_{-\infty}^{\infty} | f(x) |^2 e^{2\sigma x} \, dx \geqq e^{2\sigma a} I, \tag{4.04}$$

which is contrary to our assumption.

Conversely, if $f(x)$ vanishes for positive values of its argument, and if, for some λ,

$$\int_{-\infty}^{0} | f(x) |^2 e^{-2\lambda x} \, dx < \infty, \tag{4.05}$$

the function $F(\sigma + it)$ defined by (3.38) will belong uniformly to L_2 as a function of t for $\sigma \geqq -\lambda$. In particular, we have

THEOREM V. *The two following classes of analytic functions are identical:*

(1) *the class of all functions $F(\sigma + it)$ analytic for $\sigma > 0$, and such that*

$$\int_{-\infty}^{\infty} | F(\sigma + it) |^2 \, dt < \text{const.} \qquad [0 < \sigma < \infty]; \tag{4.06}$$

(2) *the class of all functions defined by*

$$F(\sigma + it) = \operatorname*{l.i.m.}_{A \to \infty} (2\pi)^{-1/2} \int_{-A}^{0} f(x) \, e^{x(\sigma + it)} \, dx, \tag{4.07}$$

where $f(x)$ belongs to L_2 over $(-\infty, 0)$.

We shall have

$$\operatorname*{l.i.m.}_{\sigma\to 0} F(\sigma + it) = \operatorname*{l.i.m.}_{A\to\infty} (2\pi)^{-1/2} \int_{-A}^{0} f(x)\, e^{itx}\, dx. \tag{4.08}$$

5. Theorems of the Phragmén-Lindelöf type. Let us again consider functions $F(\sigma + it)$, analytic over $-\lambda \leqq \sigma \leqq \mu$. Let us assume that

$$\int_{-\infty}^{\infty} |F(-\lambda + it)|^2\, dt < \infty, \qquad \int_{-\infty}^{\infty} |F(\mu + it)|^2\, dt < \infty, \tag{5.01}$$

and that

$$|F(\sigma + it)| < M \qquad [-\lambda \leqq \sigma \leqq \mu]. \tag{5.02}$$

In place of (3.04) we shall have for sufficiently large B

$$\begin{aligned} \left| \frac{1}{2\pi i} \int_{B}^{B+1} dA \int_{\mu+Ai}^{-\lambda+Ai} \frac{F(z)}{z - s}\, dz \right| &\leqq \frac{1}{2\pi} \int_{-\lambda}^{\mu} dz \int_{B}^{B+1} \frac{M dA}{|z + A_i - s|} \\ &\leqq \frac{M(\mu + \lambda)}{2\pi} \frac{1}{B - t}, \end{aligned} \tag{5.03}$$

so that (3.08) is established as before. Thus (3.11) is valid, and the whole argument up to (3.21) is repeated unchanged. The sole difference is that the argument this time is so turned as to *prove* that $F(\sigma + it)$ is the Fourier transform of a function $f(\sigma, x)$ belonging to L_2, instead of initially assuming it, or the equivalent fact that $F(\sigma + it)$ belongs uniformly to L_2.

It results immediately from (3.21) that $f(\sigma, t)$ belongs uniformly to L_2 over $(-\lambda, \mu)$, and hence that $F(\sigma + it)$ belongs uniformly to L_2 over this interval. We have thus proved

THEOREM VI. *If* (5.01) *and* (5.02) *are satisfied, the hypotheses and hence the conclusions of Theorems* II, III, *and* IV *are valid.*

We now appeal to the classical theorem of Phragmén and Lindelöf.* This asserts that if $F(\sigma + it)$ is analytic for $-\lambda \leqq \sigma \leqq \mu$, if

$$F(\sigma + it) = O(e^{e^{\rho t}}) \qquad [\rho < \mu + \lambda], \tag{5.04}$$

and if $F(-\lambda + it)$ and $F(\mu + it)$ are bounded, then $F(\sigma + it)$ is bounded for all t and for $-\lambda \leqq \sigma \leqq \mu$. Without any assumption further than that $F(s)$ belongs to L_2 for the ordinates at $-\lambda$ and μ, it will then follow that if $F(s)$ is analytic up to and including these ordinates and in the strip between them, and if (5.04) is fulfilled, the analytic function

$$F_\epsilon(\sigma + it) = \frac{1}{\epsilon} \int_{t}^{t+\epsilon} F(\sigma + i\tau)\, d\tau = \frac{1}{2\epsilon} \int_{\sigma+it}^{\sigma+it+i\epsilon} F(s)\, ds \tag{5.05}$$

* E. C. Titchmarsh, *Theory of Functions*, p. 178 ff.

will be bounded and will satisfy (5.04), and hence by the argument with which we have proved Theorem VI,

$$\int_{-\infty}^{\infty} |F_\epsilon(\sigma + it)|^2 \, dt \leqq \int_{-\infty}^{\infty} |F_\epsilon(-\lambda + it)|^2 \, dx + \int_{-\infty}^{\infty} |F_\epsilon(\mu + it)|^2 \, dt \qquad [-\lambda \leqq \sigma \leqq \mu]. \tag{5.06}$$

However, it is a well known theorem that if $\phi(x)$ belongs to L_2,

$$\underset{\epsilon \to 0}{\text{l.i.m.}} \frac{1}{\epsilon} \int_x^{x+\epsilon} \phi(\xi) \, d\xi = \phi(x). \tag{5.07}$$

Moreover, we have

$$\begin{aligned} \int_\alpha^\beta |F(\sigma + it)|^2 \, dt &= \lim_{\epsilon \to 0} \int_{-\infty}^{\infty} |F_\epsilon(\sigma + it)|^2 \, dt \\ &\leqq \int_{-\infty}^{\infty} |F(-\lambda + it)|^2 \, dt + \int_{-\infty}^{\infty} |F(\mu + it)|^2 \, dt, \end{aligned} \tag{5.08}$$

and hence

$$\int_{-\infty}^{\infty} |F(\sigma + it)|^2 \, dt < \text{const.} \qquad [-\lambda \leqq \sigma \leqq \mu]. \tag{5.09}$$

This yields us

THEOREM VII. *If $F(s)$ is analytic over $-\lambda \leqq \sigma \leqq \mu$, if (5.01) is satisfied, and if (5.04) is satisfied, the conclusions of Theorems II, III and IV are valid.*

We now turn our attention from the strip to the half-plane. Let $F(s)$ be analytic for $\sigma \geqq 0$, let

$$\int_{-\infty}^{\infty} |F(it)|^2 \, dt < \infty, \tag{5.10}$$

and let

$$|F(\sigma + it)| \leqq \text{const.} \qquad [0 \leqq \sigma < \infty]. \tag{5.11}$$

As before, we have

$$F(s) = \lim_{B \to \infty} \frac{1}{2\pi} \int_B^{B+1} dA \int_{-A}^{A} \left\{ \frac{F(\mu + iy)}{\mu + iy - s} - \frac{F(iy)}{iy - s} \right\} dy, \tag{5.12}$$

provided μ is large enough. This may be written

$$F(s) = \lim_{B \to \infty} \frac{1}{2\pi} \int_B^{B+1} dA \int_{-A}^{A} \frac{F(\mu + iy)}{\mu + iy - s} \, dy - \frac{1}{2\pi} \int_{-\infty}^{\infty} \frac{F(iy)}{iy - s} \, dy \tag{5.13}$$

by the argument of (3.10) and (3.11). Now if μ is large enough, and

$$\Re(s) > 0$$

we have

$$\left| \frac{1}{2\pi} \int_B^{B+1} dA \int_{-A}^{A} \left\{ \frac{F(\mu + iy)}{\mu + iy - s} - \frac{F(\mu + iy)}{\mu + iy - s_1} \right\} dy \right| \tag{5.14}$$
$$\leqq \text{const.} \int_{-\infty}^{\infty} \frac{dy}{|\mu + iy - s||\mu + iy - s_1|} = o(1) \text{ as } \mu \to \infty.$$

This is only possible if

$$F(s) = \text{const.} - \frac{1}{2\pi} \int_{-\infty}^{\infty} \frac{F(iy)}{iy - s}\, dy. \tag{5.15}$$

If we put

$$F(it) = \underset{A \to \infty}{\text{l.i.m.}}\, (2\pi)^{-1/2} \int_{-A}^{A} f(x)\, e^{itx}\, dx, \tag{5.16}$$

we obtain

$$-\frac{1}{2\pi} \int_{-\infty}^{\infty} \frac{F(iy)}{iy - s}\, dy = \frac{1}{(2\pi)^{1/2}} \int_{-\infty}^{0} f(x)\, e^{(\sigma + it)x}\, dx \tag{5.17}$$

as in (3.18). Therefore

$$-\frac{1}{2\pi} \int_{-\infty}^{\infty} \frac{F(iy)}{iy - s}\, dy \tag{5.18}$$

belongs uniformly to L_2 for $\sigma > 0$, and converges in the mean to a function of L_2 as $\sigma \to +0$. Thus the constant in (5.15) belongs to L_2 and vanishes, and we have

$$F(s) = -\frac{1}{2\pi} \int_{-\infty}^{\infty} \frac{F(iy)}{iy - s}\, dy. \tag{5.19}$$

This yields us

THEOREM VIII. *If $F(s)$ is bounded and analytic over $0 \leqq \sigma < \infty$ and if* (5.10) *is satisfied,* (4.06) *is satisfied, and we may write $F(s)$ in the form* (4.07).

If we now introduce the form of the Phragmén-Lindelöf theorem which asserts that if $F(it)$ is bounded and if uniformly for all θ in $(-\pi/2, \pi/2)$

$$\lim_{t \to \infty} \frac{1}{t} \,|\log F(te^{i\theta})\,| = 0, \tag{5.20}$$

then $F(s)$ is bounded in the half-plane $0 \leqq \sigma < \infty$, an argument quite similar to that by means of which we have proved Theorem VII will establish

THEOREM IX. *Let $F(s)$ be analytic over $0 \leqq \sigma < \infty$ and let* (5.10) *be satisfied. Let* (5.20) *be satisfied. Then* (4.06) *is satisfied, and we may write $F(s)$ in the form* (4.07).

6. **Entire functions of exponential type.** We now come to the consideration of a class of entire functions which we shall know as E. The class E consists of all entire functions $F(z)$ for which, along the real axis,

$$\int_{-\infty}^{\infty} | F(x) |^2 \, dx < \infty \tag{6.01}$$

while there exists a constant A such that

$$F(z) = o(e^{A|z|}). \tag{6.02}$$

Now, the function

$$\frac{e^{-Az}}{\epsilon} \int_{z}^{z+i\epsilon} F(iw) \, dw = G(z) \tag{6.03}$$

is bounded over the imaginary axis and the positive real axis, and is at most of exponential growth. Thus by one form of the Phragmén-Lindelöf theorem, it is bounded in the right half-plane, and satisfies the conditions laid down in the hypothesis to Theorem IX. Thus we may write

$$G(z) = \int_{-\infty}^{0} f_\epsilon(x) \, e^{zx} \, dx \tag{6.04}$$

where $f_\epsilon(x)$ belongs to L_2, or

$$\frac{1}{\epsilon} \int_{z}^{z+\epsilon} F(w) \, dw = iG(iz) \, e^{Aiz} = \int_{-\infty}^{A} i f_\epsilon(x - A) \, e^{izx} \, dx. \tag{6.05}$$

Similarly

$$\frac{1}{\epsilon} \int_{z}^{z+\epsilon} F(w) \, dw = \int_{-A}^{\infty} i g_\epsilon(x - A) \, e^{izx} \, dx. \tag{6.06}$$

That is, almost everywhere

$$\underset{B\to\infty}{\text{l. i. m.}} \frac{1}{(2\pi)^{1/2}} \int_{-B}^{B} \left[\frac{1}{\epsilon} \int_{x}^{x+\epsilon} F(y) \, dy \right] e^{-ixu} \, dx = 0 \qquad [|u| > A]. \tag{6.07}$$

Now, the Fourier transform of $F(x)$ is the limit in the mean of the Fourier transform of

$$\frac{1}{\epsilon} \int_{x}^{x+\epsilon} F(y) \, dy$$

as $\epsilon \to 0$, and hence must also vanish when $|u| > A$. That is,

$$F(z) = \int_{-A}^{A} f(u) \, e^{iuz} \, du \tag{6.08}$$

where $f(u)$ belongs to L_2 over $(-A, A)$.

On the other hand, let $f(u)$ belong to L_2 over $(-A, A)$, and let $F(z)$ be defined by (6.08). Then (6.01) and (6.02) will be satisfied for

$$\int_{-\infty}^{\infty} |F(x)|^2\,dx = \frac{1}{2\pi}\int_{-A}^{A} |f(u)|^2\,du, \tag{6.09}$$

and

$$\begin{aligned} |F(z)| &\leq \left\{\int_{-A}^{A} |f(u)|^2\,du\right\}^{1/2}\left\{\int_{-A}^{A} |e^{2iuz}|\,du\right\}^{1/2} = o\left\{\int_{-A}^{A} e^{2u|\Im(z)|}\,du\right\}^{1/2} \\ &= o(e^{u|\Im(z)|}). \end{aligned} \tag{6.10}$$

We thus obtain

THEOREM X. *The two following classes of entire functions are identical*

(1) *the class of all entire functions $F(z)$ satisfying* (6.02) *and belonging to L_2 along the real axis;*

(2) *the class of all entire functions of the form* (6.08), *where $f(x)$ belongs to L_2 over $(-A, A)$.*

An immediate corollary is

THEOREM XI. *If $F(z)$ is an entire function such that*

$$\lim_{r\to\infty} \frac{1}{r}\log^+ |F(re^{i\theta})| = 0, \tag{6.11}$$

and does not vanish identically, it cannot belong to L_2 along any line.

If it does, we may take this line to be the real axis. By Theorem X, if $F(x)$ belongs to L_2, the Fourier transform of $F(x)$ will vanish almost everywhere outside $(-A, A)$ for every $A > 0$, and will hence be equivalent to zero. Thus $F(z)$ vanishes identically, and we have a contradiction.

CHAPTER I

QUASI-ANALYTIC FUNCTIONS*

7. The problem of quasi-analytic functions. One of the chief properties of analytic functions is that they are determined over their entire range of definition by the values of all their derivatives, from the zeroth order up, at any one point. The class of analytic functions $f(x)$ of a real variable over the interval $(-1, 1)$ may be characterized by the fact that $f(x)$ is infinitely often diffenentiable and that, over this interval,

$$|f^{(\nu)}(x)|^{1/\nu} < M \qquad (\nu = 1, 2, \cdots). \tag{7.01}$$

Denjoy† showed that there are less restrictive conditions on the derivatives which likewise determine a function uniquely in terms of its value and that of its derivatives at any one point. The leading theorem in this field is that of Carleman.‡ He defines a class of functions C_A over the interval $(-1, 1)$ as follows: Let $A_0 = 1, A_1, \cdots, A_n, \cdots$ be a set of positive numbers. Then C_A is to denote the set of functions defined in the interval $(-1, 1)$, infinitely many times differentiable over that range, and satisfying the inequalities

$$\max_{-1 \leqq x \leqq 1} |f^{\nu}(x)| \leqq B^{\nu} A_{\nu} \qquad (\nu = 0, 1, 2, \cdots), \tag{7.02}$$

where B is a constant which may depend on $f(x)$. We say that the class C_A is *quasi-analytic* if a function of C_A is defined completely over $(-1, 1)$ by the values of its derivatives $f^{(\nu)}(x)$ $(\nu = 0, 1, 2, \cdots)$ at a single point x_0, or, what is the same thing, if the equations

$$f^{(\nu)}(x_0) = 0 \qquad (\nu = 0, 1, 2, \cdots), \tag{7.03}$$

together with the condition that $f(x)$ belongs to C_A, imply that $f(x)$ vanishes identically. The theorem is the following:

CARLEMAN'S THEOREM. *A necessary and sufficient condition that C_A should be quasi-analytic is that the integral*

$$\int_0^{\infty} \log\left(\sum_{\nu=0}^{\infty} \frac{x^{2\nu}}{A_\nu^2}\right) \frac{dx}{1+x^2} \tag{7.04}$$

* Cf. Paley and Wiener, *Notes on the theory and application of Fourier transforms*, Note I, *On quasi-analytic functions*, Transactions of the American Mathematical Society, vol. 35, pp. 348–353.

† M. Denjoy, Comptes Rendus, vol. 173 (1921), p. 1329.

‡ T. Carleman, *Les Fonctions Quasi-Analytiques*, Paris, 1926.

should diverge, or what is the same thing, that the least non-increasing majorant of the series

$$\sum_{\nu=0}^{\infty} 1/(A_\nu)^{1/\nu} \tag{7.05}$$

should diverge.

The equivalence of the two conditions has been established by Carleman in his book.* Here we shall make no use of the second condition. In this chapter we shall give a proof of Carleman's Theorem for the class C_A and also of a similar theorem for other closely related classes.

We wish to define quasi-analyticity in the first instance for another class of functions. This definition involves the integrals of the squares of the moduli of the derivatives rather than the maxima of these moduli. Defining

$$A_0 = 1, A_1, \cdots, A_n, \cdots$$

as before as a set of positive numbers, we shall take C'_A to be the set of those functions defined in the interval $(-1, 1)$, infinitely many times differentiable over that range, and satisfying the inequalities

$$\int_{-1}^{1} |f^{(\nu)}(x)|^2 \, dx \leqq B^\nu A_\nu^2 \qquad (\nu = 0, 1, 2, \cdots), \tag{7.06}$$

where B is a constant which may depend on $f(x)$. Clearly C_A will be contained in C'_A.

On the other hand, let

$$\int_{-1}^{1} |f^{(\nu)}(x)|^2 \, dx \leqq B^\nu F_\nu^2 \qquad (\nu = 0, 1, 2, \cdots), \tag{7.07}$$

and for some ξ in $(-1, 1)$, let

$$f^{(\nu)}(\xi) = 0 \qquad (\nu = 0, 1, 2, \cdots). \tag{7.071}$$

Then, by the Schwarz inequality, if x lies in $(-1, 1)$,

$$|f^{(\nu)}(x)|^2 = \left| \int_x^\xi f^{(\nu+1)}(y) \, dy \right|^2 \leqq 2 \int_{-1}^{1} |f^{(\nu+1)}(y)|^2 \, dy \leqq 2B^{\nu+1} F_{\nu+1}^2. \tag{7.072}$$

It follows that there exists a C such that

$$|f^{(\nu)}(x)| \leqq C^\nu F_{\nu+1}. \tag{7.08}$$

Now let us put

$$A_\nu = F_{\nu+1} \qquad (\nu = 0, 1, 2, \cdots). \tag{7.081}$$

* Carleman, loc. cit., pp. 50 ff.

It will then follow that C'_F will be contained in C_A as far as functions are concerned which vanish with all their derivatives at some point. We have

$$
\begin{aligned}
\int_1^\infty \log\left(\sum_{\nu=0}^\infty \frac{x^{2\nu}}{F_\nu^2}\right)\frac{dx}{1+x^2} &= \int_1^\infty \log\left(1+\sum_{\nu=0}^\infty \frac{x^{2(\nu+1)}}{A_\nu^2}\right)\frac{dx}{1+x^2} \\
&= \int_1^\infty \log\left(\frac{1}{x^2}+\sum_{\nu=0}^\infty \frac{x^{2\nu}}{A_\nu^2}\right)\frac{dx}{1+x^2} + \int_1^\infty \frac{\log x^2}{1+x^2}\,dx \qquad (7.09)\\
&< \int_1^\infty \log\left(\sum_{\nu=0}^\infty \frac{x^{2\nu}}{A_\nu^2}\right)\frac{dx}{1+x^2} + \text{const.}
\end{aligned}
$$

Also,

$$
\int_1^\infty \log\left(1+\sum_{\nu=0}^\infty \frac{x^{2(\nu+1)}}{A_\nu^2}\right)\frac{dx}{1+x^2} > \int_1^\infty \log\sum_{\nu=0}^\infty \left(\frac{x^{2\nu}}{A_\nu^2}\right)\frac{dx}{1+x^2}. \qquad (7.10)
$$

Therefore

$$
\begin{aligned}
\int^\infty \log\left(\sum_{\nu=0}^\infty \frac{x^{2\nu}}{A_\nu^2}\right)\frac{dx}{1+x^2} &< \int_1^\infty \left(\log\sum_{\nu=0}^\infty \frac{x^{2\nu}}{F_\nu^2}\right)\frac{dx}{1+x^2} \qquad (7.11)\\
&< \int_1^\infty \log\left(\sum_{\nu=0}^\infty \frac{x^{2\nu}}{A_\nu^2}\right)\frac{dx}{1+x^2} + \text{const.}
\end{aligned}
$$

Thus the integrals (7.04) determined by A_ν and F_ν converge or diverge together, since the part of the integrals from 0 to 1 determines neither convergence nor divergence.

The class, C'_A, may of course be transferred to an infinite range by the substitution

$$
f_1(y) = \frac{f(\tan^{-1} y)}{(1+y^2)^{1/2}} \qquad (7.12)
$$

which transforms a function $f(x)$ belonging to L_2 over the range $(-1, 1)$ into a function $f_1(y)$ belonging to L_2 over $(-\infty, \infty)$. Besides this direct and rather brutal transference of the theorem, it is possible to form an analogous theorem in which the interval $(-1, 1)$ used to define the class C'_A is replaced bodily by the *infinite interval* $(-\infty, \infty)$, and the inequality (7.06) becomes

$$
\int_{-\infty}^\infty |f^{(\nu)}(x)|^2\,dx \leqq B^\nu A_\nu^2 \qquad (\nu = 0, 1, 2, \cdots). \qquad (7.13)
$$

This theorem over an infinite range is true, and may be deduced from the following fundamental

THEOREM XII. *Let $\phi(x)$ be a real non-negative function not equivalent to zero, defined for $-\infty < x < \infty$, and of integrable square in this range. A necessary and* *sufficient condition that there should exist a real- or complex-valued function $F(x)$ defined in the same range, vanishing for $x \geqq x_0$ for some number x_0, and such that the Fourier transform $G(x)$ of $F(x)$ should satisfy $| G(x) | = \phi(x)$, is that*

$$\int_{-\infty}^{\infty} \frac{| \log \phi(x) |}{1 + x^2} dx < \infty. \tag{7.14}$$

The relevance of this theorem to the theory of quasi-analytic functions depends on the following facts: (1) a condition of boundedness on the Fourier transform of a function is closely related to a condition of boundedness on its derivatives; (2) a function $F(x)$ vanishing over a half-line but not identically vanishing cannot be determined by all its derivatives at any point, and is typical of the class of non-quasi-analytic functions. Theorem XII, while similar to one of de la Vallée-Poussin,* is much more definite in that, while he concerns himself with the order of magnitude of the coefficients of a Fourier series of a function vanishing with all its derivatives at some fixed point, we undertake actually to fix the modulus of the transform of a function of the type, subject of course to the convergence of (7.14).

8. **Proof of the fundamental theorem on quasi-analytic functions.** We turn to the proof of Theorem XII. Let us suppose first that the integral (7.14) converges. We write for $z = x + iy$, $y > 0$,

$$\lambda(z) = \frac{1}{\pi} \int_{-\infty}^{\infty} \frac{\log \phi(x') y}{(x - x')^2 + y^2} dx', \tag{8.01}$$

which is harmonic in the half-plane $y > 0$. Let $\mu(z)$ be its conjugate, and write

$$h(z) = \exp (\lambda(z) + i\mu(z)). \tag{8.02}$$

It is well known, by an argument of the Fatou† type, that for almost all values of x,

$$\lim_{y \to 0} \lambda(x + iy) = \log \phi(x), \tag{8.03}$$

or, what is the same thing,

$$\lim_{y \to 0} | h(x + iy) | = \phi(x). \tag{8.04}$$

The geometric mean of two positive quantities cannot exceed the arithmetic mean. By this property, extended to integrals, or in other words, by the convexity property of the logarithm,

$$| h(x + iy) | = e^{\lambda(z)} \leqq \frac{1}{\pi} \int_{-\infty}^{\infty} \frac{\phi(x') y}{(x - x')^2 + y^2} dx'. \tag{8.05}$$

* Carleman, loc. cit., pp. 76 and 91.

† E. Landau, *Darstellung und Begründung einiger neuerer Ergebnisse der Funktionentheorie*, 1929, p. 40.

Hence, by the Schwarz inequality,

$$
\begin{aligned}
\int_{-\infty}^{\infty} | h(x+iy) |^2 \, dx &\leqq \frac{1}{\pi^2}\int_{-\infty}^{\infty} dx \int_{-\infty}^{\infty} \frac{[\phi(x')]^2\, y}{(x-x')^2+y^2}\, dx' \int_{-\infty}^{\infty} \frac{y\, dx''}{(x-x'')^2+y^2} \\
&= \frac{1}{\pi}\int_{-\infty}^{\infty} dx \int_{-\infty}^{\infty} \frac{[\phi(x')]^2\, y}{(x-x')^2+y^2}\, dx' \\
&= \frac{1}{\pi}\int_{-\infty}^{\infty} [\phi(x')]^2\, dx' \int_{-\infty}^{\infty} \frac{y\, dx}{(x-x')^2+y^2} \\
&= \int_{-\infty}^{\infty} [\phi(x')]^2\, dx'.
\end{aligned} \tag{8.06}
$$

That is, the function $h(is)$ satisfies condition (1) of Theorem V, and we may apply that theorem. This gives us

$$
h(x+iy) = \underset{A\to\infty}{\text{l. i. m.}}\ (2\pi)^{-1/2} \int_{-A}^{A} F(\xi)\, e^{-i\xi(x+iy)}\, d\xi, \tag{8.07}
$$

where $F(x)$ vanishes for $x \geqq 0$, and belongs to L_2. On the other hand, it follows from (8.07) that

$$
\underset{y\to+0}{\text{l. i. m.}}\ h(x+iy) = \underset{A\to\infty}{\text{l. i. m.}}\ (2\pi)^{-1/2} \int_{-A}^{A} F(\xi)\, e^{-i\xi x}\, d\xi. \tag{8.08}
$$

Let us put

$$
\underset{y\to 0}{\text{l. i. m.}}\ h(x+iy) = G(x). \tag{8.09}
$$

By (8.04), we have almost everywhere

$$
| G(x) | = \phi(x), \tag{8.10}
$$

and one part of Theorem XII is proved.

We still have to prove that if $G(x)$ is given as in Theorem XII, (7.14) follows. We suppose that $F(x')$ vanishes for $x' > x_0$, and we may suppose without loss of generality that $x_0 = 0$. Let us write

$$
\begin{cases}
G(x) = \underset{N\to\infty}{\text{l. i. m.}}\ (2\pi)^{-1/2} \displaystyle\int_{-N}^{N} F(x')\, e^{-ixx'}\, dx'; \\
\psi(z) = (2\pi)^{-1/2} \displaystyle\int_{-\infty}^{\infty} F(x')\, e^{-izx'}\, dx', \qquad \Im(z) > 0;
\end{cases} \tag{8.11}
$$

where the second integral is taken along a horizontal line in the z-plane. The function $\psi(z)$ is readily seen to be an analytic function in the half-plane $\Im(z) > 0$. Suppose that we map the half-plane $\Im(z) > 0$ into the circle $|\zeta| < 1$ ($\zeta = re^{i\theta}$), by means of $z = i(\zeta+1)/(\zeta-1)$, and that $G(x)$ becomes $\Gamma(e^{i\theta})$ and $\psi(z)$ becomes $\gamma(\zeta)$. Then it is easily seen that

$$
\int_{-\pi}^{\pi} | \Gamma(e^{i\theta}) |^2 \, d\theta = 2 \int_{-\infty}^{\infty} \frac{| G(x) |^2}{1+x^2}\, dx, \tag{8.12}
$$

so that Γ certainly is of class L_2. Also a simple computation shows that if $re^{i\phi}$ is the image of $x' + iy'$, then

$$(2\pi)^{-1} \int_{-\pi}^{\pi} \Gamma(e^{i\theta}) \frac{1 - r^2}{1 - 2r \cos (\theta - \phi) + r^2} d\theta = \pi^{-1} \int_{-\infty}^{\infty} G(x) \frac{y' \, dx}{(x - x')^2 + y'^2}$$

$$= \pi^{-1} \int_{-\infty}^{\infty} \frac{y' \, dx}{(x - x')^2 + y'^2} \operatorname*{l.i.m.}_{N\to\infty} (2\pi)^{-1/2} \int_{-N}^{0} F(\xi)\, e^{-ix\xi}\, d\xi$$

$$= \lim_{N\to\infty} \pi^{-1} \int_{-\infty}^{\infty} \frac{y' \, dx}{(x - x')^2 + y'^2} (2\pi)^{-1/2} \int_{-N}^{0} F(\xi)\, e^{-ix\xi}\, d\xi$$

$$= \lim_{N\to\infty} \pi^{-1} \int_{-N}^{0} \frac{F(\xi)\, d\xi}{(2\pi)^{1/2}} \int_{-\infty}^{\infty} \frac{e^{-ix\xi}\, y' \, dx}{(x - x')^2 + y'^2}$$

$$= \lim_{N\to\infty} (2\pi)^{-1/2} \int_{-N}^{0} F(\xi)\, e^{-ix'\xi + y'\xi}\, d\xi = \psi(x' + iy') = \gamma(re^{i\phi}),$$

so that γ is in fact the Poisson integral of $\Gamma(e^{i\theta})$. Then

$$(8.14) \qquad (2\pi)^{-1} \int_{-\pi}^{\pi} \log^+ | \gamma(re^{i\theta}) | \, d\theta \leqq (2\pi)^{-1} \int_{-\pi}^{\pi} | \gamma(re^{i\theta}) |^2 \, d\theta \leqq (2\pi)^{-1} \int_{-\pi}^{\pi} | \Gamma(e^{i\theta}) |^2 \, d\theta.$$

We now invoke one of the most important theorems in analysis: Jensen's theorem.* This states

THEOREM XIII. *Let $f(z)$ be analytic for $| z | < R$. Suppose that $f(0)$ is not zero, and let $r_1, r_2, \cdots, r_n, \cdots$ be the moduli of the zeros of $f(z)$ in the circle $| z | < R$, arranged in a non-decreasing sequence. Then if $r_n \leqq r \leqq r_{n+1}$,*

$$(8.15) \qquad \log \frac{r^n \, | f(0) |}{r_1 r_2 \cdots r_n} = \frac{1}{2\pi} \int_0^{2\pi} \log | f(re^{i\theta}) | \, d\theta,$$

where every zero is counted the number of times of its multiplicity.

It follows that if $\gamma(0) \neq 0$,

$$(8.16) \qquad \frac{1}{2\pi} \int_{-\pi}^{\pi} \log | \gamma(re^{i\theta}) | \, d\theta \geqq \log | \gamma(0) |.$$

Furthermore,

$$(8.17) \qquad \frac{1}{2\pi} \int_{-\pi}^{\pi} \log | \gamma(re^{i\theta}) | \, d\theta = \frac{1}{2\pi} \int_{-\pi}^{\pi} \log^+ | \gamma(re^{i\theta}) | \, d\theta + \frac{1}{2\pi} \int_{-\pi}^{\pi} \log^- | \gamma(re^{i\theta}) | \, d\theta,$$

* E. C. Titchmarsh, *Theory of Functions*, p. 125.

and

$$(8.18)\qquad \begin{aligned}\frac{1}{2\pi}\int_{-\pi}^{\pi} |\log|\gamma(re^{i\theta})||\,d\theta &= \frac{1}{2\pi}\int_{-\pi}^{\pi}\log^+|\gamma(re^{i\theta})|\,d\theta - \frac{1}{2\pi}\int_{-\pi}^{\pi}\log^-|\gamma(re^{i\theta})|\,d\theta \\ &= \frac{1}{\pi}\int_{-\pi}^{\pi}\log^+|\gamma(re^{i\theta})|\,d\theta - \frac{1}{2\pi}\int_{-\pi}^{\pi}\log|\gamma(re^{i\theta})|\,d\theta.\end{aligned}$$

That is, by (8.14) and (8.16), we have uniformly in r

$$(8.19)\qquad \frac{1}{2\pi}\int_{-\pi}^{\pi}|\log|\gamma(re^{i\theta})||\,d\theta \leqq \frac{1}{\pi}\int_{-\pi}^{\pi}|\Gamma(e^{i\theta})|^2\,d\theta - \log|\gamma(0)|.$$

In any case, if m is the multiplicity of 0 as a root of γ,

$$(8.191)\qquad \begin{aligned}\frac{1}{2\pi}\int_{-\pi}^{\pi}|\log|\gamma(re^{i\theta})||\,d\theta &\leqq \frac{1}{2\pi}\int_{-\pi}^{\pi}\left|\log\left|\frac{\gamma(re^{i\theta})}{r^m e^{mi\theta}}\right|\right|d\theta - m\log r \\ &\leqq \frac{1}{\pi}\int_{-\pi}^{\pi}|\Gamma(e^{i\theta})|^2\,d\theta - \log\left|\frac{\gamma(\zeta)}{\zeta^m}\right|_{\zeta=0} - m\log r.\end{aligned}$$

Since finally $\log|\gamma(re^{i\theta})|$ tends almost everywhere to $\log|\Gamma(e^{i\theta})|$ as $r \to 1$, we have

$$(8.20)\qquad \frac{1}{2\pi}\int_{-\pi}^{\pi}|\log|\Gamma(e^{i\theta})||\,d\theta < \infty,$$

and inverting again to the half-plane, this implies that

$$(8.21)\qquad \int_{-\infty}^{\infty}\frac{|\log|G(x)||}{1+x^2}\,dx < \infty.$$

9. **Proof of Carleman's theorem.** Let integral (7.04) converge, and let

$$(9.01)\qquad [\phi(x)]^2 = 10^{-1}(1+x^2)^{-1}\left[\sum_{\nu=0}^{\infty}\frac{x^{2\nu}}{A_\nu^2}\right]^{-1}.$$

Clearly

$$(9.02)\qquad \int_{-\infty}^{\infty}\frac{|\log\phi(x)|}{1+x^2}\,dx < \infty, \qquad \int_{-\infty}^{\infty}[\phi(x)]^2\,dx < \infty.$$

Thus by Theorem XII there exists a function $F(x)$ belonging to L_2, vanishing for $x > 0$ but not equivalent to zero, and with a Fourier transform $G(x)$ satisfying $|G(x)| = \phi(x)$. Furthermore,

$$(9.03)\qquad \begin{aligned}\int_{-\infty}^{\infty}|F^{(\nu)}(x)|^2\,dx &= \int_{-\infty}^{\infty}|G(x)|^2 x^{2\nu}dx = \int_{-\infty}^{\infty}[\phi(x)]^2 x^{2\nu}\,dx \\ &\leqq \int_{-\infty}^{\infty}[10(1+x^2)]^{-1}\left(\frac{x^{2\nu}}{A_\nu^2}\right)^{-1}x^{2\nu}\,dx \leqq A_\nu^2.\end{aligned}$$

Thus the divergence of the first integral (9.02) is certainly necessary for the quasi-analyticity of the class C_A' over the infinite line, and a fortiori over $(-1,$

1). However, given a sequence A_ν for which integral (7.04) converges, then from (7.081) we determine a sequence F_ν which also has the integral corresponding to (7.04) convergent. By (9.03), a function exists belonging to C'_F and vanishing with all its derivatives at the origin. By (7.08), this function will also be contained in C_A. Thus the divergence of the integral (7.04) is necessary for the quasi-analyticity of C_A.

We now come to the question of the sufficiency of Carleman's condition for quasi-analyticity, and we shall first consider the case of C'_A over the infinite interval. Let $f(x)$ be a function which is not identically zero, but which vanishes with all its derivatives at $x = 0$, and is everywhere infinitely often differentiable. We wish to show that integral (7.04) converges for every class C'_A to which $f(x)$ belongs. Let $F(x)$ be identical with $f(x)$ for negative x, and let it vanish for all positive x. Let $G(x)$ be the Fourier transform of $F(x)$. In formula (7.13) let $B = 1$. This is no real restriction. We obtain

$$(9.04)\quad A_\nu^2 \geqq \int_{-\infty}^{\infty} |f^{(\nu)}(x)|^2\,dx \geqq \int_{-\infty}^{\infty} |F^{(\nu)}(x)|^2\,dx = \int_{-\infty}^{\infty} |G(x)|^2 x^{2\nu}\,dx.$$

It follows that

$$(9.05)\quad \begin{aligned} \log\left(\sum_{\nu=0}^{\infty}\frac{r^{2\nu}}{A_\nu^2}\right) &\leqq \log\left(\sum_{\nu=0}^{\infty}\left[\int_{-\infty}^{\infty}|G(x)|^2\left(\frac{x}{r}\right)^{2\nu}dx\right]^{-1}\right)\\ &\leqq \log\left(\sum_{\nu=0}^{\infty}\left[\int_{2r}^{2r+1}|G(x)|^2\left(\frac{x}{r}\right)^{2\nu}dx\right]^{-1}\right)\\ &\leqq \log\left(2\left[\int_{2r}^{2r+1}|G(x)|^2\,dx\right]^{-1}\right)\\ &\leqq 2\int_{2r}^{2r+1}|\log(2^{-1/2}|G(x)|)|\,dx. \end{aligned}$$

Hence

$$(9.06)\quad \begin{aligned} \int_1^{\infty}\log\left(\sum_{\nu=0}^{\infty}\frac{r^{2\nu}}{A_\nu^2}\right)\frac{dr}{r^2} &\leqq 2\int_1^{\infty}\frac{dr}{r^2}\int_{2r}^{2r+1}|\log(2^{-1/2}|G(x)|)|\,dx\\ &\leqq 2\int_2^{\infty}|\log(2^{-1/2}|G(x)|)|\,dx\int_{x/2-1/2}^{x/2}\frac{dr}{r^2}\\ &\leqq 20\int_2^{\infty}|\log(2^{-1/2}|G(x)|)|\,x^{-2}\,dx < \infty. \end{aligned}$$

Thus the divergence of (7.04) is sufficient for quasi-analyticity over the infinite interval. This proves the analogue of Carleman's theorem for the class C_A over the infinite interval.

We now turn to the case of the interval $(-1, 1)$. Here we require the following lemma of Mr. N. Levinson:

Let $f(x)$ be of class C_A over $(-1, 1)$ and vanish over $(-1, \xi)$ where $\xi \geqq 0$. Then if $F(t) = e^{-t} f(1 - e^{-t})$ for $t \geqq 0$, and $F(t) = 0$ for $t < 0$, $F(t)$ is of class C'_A over $(-\infty, \infty)$.

For, for $t \geqq 0$;

$$F(t) = e^{-t} f(1 - e^{-t});$$

$$\frac{d}{dt} F(t) = e^{-2t} f'(1 - e^{-t}) - e^{-t} f(1 - e^{-t});$$

$$\frac{d^2}{dt^2} F(t) = e^{-3t} f''(1 - e^{-t}) - 3\, e^{-2t} f'(1 - e^{-t}) + e^{-t} f(1 - e^{-t});$$

(9.061) ....

$$\frac{d^n}{dt^n} F(t) = e^{-(n+1)t} f^{(n)}(1 - e^{-t}) + a_1\, e^{-nt} f^{(n-1)}(1 - e^{-t}) + \cdots + a_n\, e^{-t} f(1 - e^{-t});$$

$$\frac{d^{n+1}}{dt^{n+1}} F(t) = e^{-(n+2)t} f^{(n+1)}(1 - e^{-t}) + b_1\, e^{-(n+1)t} f^{(n)}(1 - e^{-t}) + \cdots + b_{n+1}\, e^{-t} f(1 - e^{-t});$$

....

Here $f^{(k)}(1 - e^{-t}) = f^{(k)}(x)$ where $x = 1 - e^{-t}$. We have above

$$b_1 = -(n+1) + a_1, \quad b_2 = -n a_1 + a_2, \cdots, \quad b_n = -2\, a_{n-1} + a_n, \quad b_{n+1} = -a_n. \tag{9.062}$$

From this we get

$$| b_\alpha | \leqq (n+1)\, | a_{\alpha-1} | + | a_\alpha | \qquad (\alpha = 1, 2, \cdots, n+1); \tag{9.063}$$

where $a_0 = 1$ and $a_{n+1} = 0$. We shall use the principle of induction to obtain inequalities involving a_α. Let us assume that

$$| a_\alpha | \leqq \frac{n^{2\alpha}}{\alpha\,!} 2^n \qquad (\alpha = 1, 2, \cdots, n). \tag{9.064}$$

Then

$$| b_\alpha | \leqq \left[(n+1) \frac{n^{2\alpha-2}}{(\alpha-1)\,!} + \frac{n^{2\alpha}}{\alpha\,!} \right] 2^n < \frac{(n+1)^{2\alpha}}{\alpha\,!} 2^{n+1} \qquad (\alpha = 1, 2, \cdots, n+1). \tag{9.065}$$

But the assumed inequality holds for $n = 1$ and we have just shown that if it holds for n it holds for $n + 1$, and (9.064) is proved for all values of n. Thus for $t \geqq 0$,

$$\left| \frac{d^n}{dt^n} F(t) \right| \leqq 2^n \left[e^{-(n+1)t} \left| f^{(n)}(1 - e^{-t}) \right| + \frac{n^2}{1!} e^{-nt} \left| f^{(n-1)}(1 - e^{-t}) \right| + \cdots + \frac{n^{2n}}{n!} e^{-t} \left| f(1 - e^{-t}) \right| \right]. \tag{9.066}$$

Now since $f(x)$ vanishes with all its derivatives at $x = 0$,

$$f^{(n-\alpha)}(x) = \frac{1}{(\alpha - 1)!} \int_0^x (x - \xi)^{\alpha-1} f^{(n)}(\xi)\, d\xi \qquad (\alpha > 0). \tag{9.07}$$

Therefore

$$| f^{(n-\alpha)}(x) | \leqq \frac{\max | f^n(\xi) |}{(\alpha - 1)!}. \tag{9.08}$$

Since $f(x)$ belongs to C_A it follows from (7.02) that

$$| f^{(n-\alpha)}(x) | \leqq \frac{B^n A_n}{(\alpha - 1)!}. \tag{9.081}$$

Using this and the fact that $e^{-t} \leqq 1$ for $t \geqq 0$, we have

$$\begin{aligned} \left| \frac{d^n}{dt^n} F(t) \right| &\leqq e^{-t} (2B)^n A_n \left[1 + n^2 + \frac{n^4}{2!} + \frac{n^6}{2!\,3!} + \cdots + \frac{n^{2n}}{(n-1)!\,n!} \right] \\ &< e^{-t} (2B)^n A_n\, n \left[1 + n^2 + \frac{n^4}{(2!)^2} + \frac{n^6}{(3!)^2} + \cdots + \frac{n^{2n}}{(n!)^2} \right]. \end{aligned} \tag{9.09}$$

If we select a certain set of terms of

$$e^{2n} = \left[1 + n + \frac{n^2}{2!} + \frac{n^3}{3!} + \cdots \right]^2$$

we see that

$$\left| \frac{d^n}{dt^n} F(t) \right| < e^{-t} (2B)^n A_n\, n\, e^{2n} < e^{-t} B^n A_n\, e^{3n}. \tag{9.10}$$

If $C = 2Be^3$ then

$$\left| \frac{d^n}{dt^n} F(t) \right| < e^{-t} C^n A_n \text{ for } t \geqq 0. \tag{9.11}$$

Hence

$$\int_{-\infty}^{\infty} \left| \frac{d^n F(t)}{dt^n} \right|^2 dt < C^{2n} A_n^2 \int_0^{\infty} e^{-2t}\, dt < C^{2n} A_n^2. \tag{9.12}$$

Thus $F(t)$ belongs to C_A' over $(-\infty, \infty)$. This proves the lemma.

Now let $f(x)$ be of class C_A over $(-1, 1)$ and let the sequence, A_n, be such that (7.04) diverges. We shall show that if $f(x)$ and all its derivatives vanish at some point in the interval, then it must vanish identically, which will prove the sufficiency of the divergence of (7.04) for the quasi-analyticity of C_A. For, let us assume that $f(x)$ does not vanish identically. Then, by translation, or by translation and the substitution $y = -x$, it is always possible to get a function $f_1(x)$ such that the function and all its derivatives vanish for $x = \xi$ (where $0 \leqq \xi < 1$), such that $f_1(x)$ still belongs to C_A, and such that $f_1(x)$ is not equivalent to zero over $1 > x > \xi$.

We now define $f_2(x) = 0$ for $-1 \leqq x < \xi$ and $f_2(x) = f_1(x)$ for $1 > x > \xi$. But $f_2(x)$ satisfies the conditions of the preceding lemma. From this lemma and from the sufficiency of the divergence of (7.04) for quasi-analyticity of C'_A over the infinite interval, which has already been shown, it follows that $F(t)$ of the lemma and therefore $f_2(x)$ must vanish identically, which leads to a contradiction. Thus the divergence of (7.04) is also sufficient for the quasi-analyticity of C_A over $(-1, 1)$. This completes the proof of Carleman's theorem for the class C_A.

The divergence of (7.04) is also sufficient for the class C'_F over the interval $(-1, 1)$. For, assume that this is not true; then a function exists vanishing with all its derivatives at some point in the closed interval $(-1, 1)$, but not vanishing identically, belonging to C'_F, and such that (7.04) diverges. From (7.08) this function also belongs to a class C_A for which (7.04) diverges, and from Carleman's Theorem which we have just proved for the classes C_A, it follows that the function vanishes identically. This completes the sufficiency proof. The necessity of the condition has already been shown for the class C'_A. Thus we have proved the analogue of the Carleman Theorem for the classes C'_A over $(-1, 1)$.

10. **The modulus of the Fourier transform of a function vanishing for large arguments.** Let $f(x)$ be a function belonging to L over $(-A, A)$, and vanishing outside $(-A, A)$. This statement is equivalent to the assertion that it is the product of two functions of L_2, one of which vanishes to the left of $-A$, while the other vanishes to the right of A. Let these functions be $f_1(x)$ and $f_2(x)$, and let their Fourier transforms be $g_1(u)$ and $g_2(u)$. Then by the Parseval theorem, the Fourier transform of $f(x)$ is

$$g(u) = (2\pi)^{-1/2} \int_{-\infty}^{\infty} g_1(v)\, g_2(u - v)\, dv. \tag{10.01}$$

Furthermore, by Theorem XII, the real functions $| g_1(u) |$ and $| g_2(u) |$ are subject to the sole condition that

$$\int_{-\infty}^{\infty} \frac{| \log | g_1(u) ||}{1 + u^2}\, du < \infty, \qquad \int_{-\infty}^{\infty} \frac{| \log | g_2(u) ||}{1 + u^2}\, du < \infty. \tag{10.02}$$

This condition is manifestly independent of A. Moreover, without changing $| g_1(u) |$ and $| g_2(u) |$, it is possible so to choose f_1 and f_2 that $f_1(x)$ and $f_2(x)$ have a product which is not equivalent to zero, without any further assumption than (10.02).

To see this, let $h_1(x)$ and $h_2(x)$ be functions which are equal respectively to $|f_1(x)|$ and $|f_2(x)|$ over $(-A, A)$ and zero elsewhere. We may certainly so choose f_1 and f_2 that h_1 and h_2 are not null. Then if

$$h_1(x) \sim \sum_{-\infty}^{\infty} a_n\, e^{in\pi x/(2A)}, \qquad h_2(-x) \sim \sum_{-\infty}^{\infty} b_n\, e^{in\pi x/(2A)} \qquad (-2A \leqq x \leqq 2A), \tag{10.03}$$

we have

$$\int_{-2A}^{2A} h_1(y)\, h_2(-x+y)\, dy = 4A \sum_{-\infty}^{\infty} a_n\, b_n\, e^{in\pi x/(2A)}, \tag{10.04}$$

and since a_0 and b_0 are clearly not equal to zero, we may find an x_0, which must lie within $(-A, A)$, such that

$$\int_{-2A}^{2A} h_1(y)\, h_2(-x_0+y)\, dy \neq 0. \tag{10.05}$$

It will immediately result that

$$f_1(x)\, f_2(-x_0+x) \not\equiv 0, \tag{10.06}$$

and if we replace $f_2(x)$ by $f_2(x - x_0)$, we obtain values of f_1 and f_2 corresponding to the given values of $|g_1|$ and $|g_2|$ for which $f_1 f_2$ is not equivalent to zero.

Now, by (10.01),

$$\begin{aligned} |g(u)| &\leqq (2\pi)^{-1/2} \int_{-\infty}^{\infty} |g_1(v)|\,|g_2(u-v)|\, dv \\ &\leqq (2\pi)^{-1/2} \int_{-\infty}^{u/2} |g_1(v)|\,|g_2(u-v)|\, dv \\ &\quad + (2\pi)^{-1/2} \int_{u/2}^{\infty} |g_1(v)|\,|g_2(u-v)|\, dv \\ &\leqq (2\pi)^{-1/2} \int_{-\infty}^{\infty} |g_1(v)|\, dv \limsup_{w \geqq u/2} |g_2(w)| \\ &\quad + (2\pi)^{-1/2} \int_{-\infty}^{\infty} |g_2(v)|\, dv \limsup_{w \geqq u/2} |g_1(w)|. \end{aligned} \tag{10.07}$$

If there exists some function $\phi(u) \geqq 0$ which is monotonely decreasing for $|u| \to \infty$ and which satisfies

$$A\,|g_i(u)| \leqq \phi(u) \leqq B\,|g_i(u)| \qquad (i = 1, 2;\ B \geqq A > 0), \tag{10.08}$$

then

$$\int_{-\infty}^{\infty} \frac{|\log|g(u)||}{1+u^2}\, du < \infty \tag{10.09}$$

implies (10.02).

CHAPTER II

SZÁSZ'S THEOREM

11. **Certain theorems of closure.** In the present chapter, we wish to investigate the closure over (a, b) of certain assemblages of functions $\{f_n(x)\}$, $f_n(x) \subset L_2$. The set of functions $\{f_n(x)\}$ is said to be *closed over* (a, b) if

$$\int_b^a f(x) f_n(x) \, dx = 0 \qquad (n = 1, 2, \cdots) \tag{11.01}$$

implies that $f(x)$ vanishes except over a set of zero measure if $f(x) \subset L_2$. The set of functions $\{f_n(x)\}$ is said to be *complete* if to any function $f(x)$ belonging to L_2 and any positive ϵ there is a polynomial

$$P_n(x) = \sum_1^n a_k f_k(x) \tag{11.02}$$

such that

$$\int_a^b | P_n(x) - f(x) |^2 \, dx < \epsilon. \tag{11.021}$$

We shall now prove a theorem that shows the significance of the closure of a sequence of functions. This theorem states that *a set of functions is closed when, and only when, it is complete.*

First we shall prove the classical theorem for a set $\{\phi_n(x)\}$ which is normal and orthogonal: that is, for which

$$\int_a^b \phi_n(x) \overline{\phi_m(x)} \, dx = 0 \qquad (m \neq n); \tag{11.03}$$

$$\int_a^b \phi_n(x) \overline{\phi_n(x)} \, dx = 1 \qquad (n = 1, 2, \cdots). \tag{11.031}$$

Now let $f(x)$ be any function belonging to L_2. Then

$$\begin{aligned} \int_a^b \left| f(x) - \sum_1^n a_k \phi_k(x) \right|^2 dx &= \int_a^b \left(f(x) - \sum_1^n a_k \phi_k(x) \right) \\ &\qquad \times \left(\overline{f(x)} - \sum_1^n \bar{a}_k \overline{\phi_k(x)} \right) dx \\ &= \int_a^b | f^2 | \, dx - \sum_1^n \left| \int_a^b f \bar{\phi}_k \, dx \right|^2 + \sum_1^n \left| a_k - \int_a^b f \bar{\phi}_k \, dx \right|^2 . \end{aligned} \tag{11.04}$$

This expression is positive for all choices of a_k. Therefore it is a minimum for

$$a_k = \int_a^b f\bar{\phi}_k \, dx,$$

which causes the last term to vanish. If we also let $n \to \infty$, we have

$$\int_a^b |f|^2 \, dx - \sum_1^\infty \left| \int_a^b f\bar{\phi}_k \, dx \right|^2 \geqq 0. \tag{11.041}$$

This is known as the Bessel inequality.

If the set $\{\phi_n(x)\}$ is known to be complete, then from the definition of completeness and the minimum property of

$$a_k = \int_a^b f\bar{\phi}_k \, dx,$$

we see that the Bessel inequality becomes an equality, and we have

$$\int_a^b |f|^2 \, dx = \sum_1^\infty \left| \int_a^b f\bar{\phi}_k \, dx \right|^2. \tag{11.042}$$

Thus no function not equivalent to zero can exist which is orthogonal to every $\phi_k(x)$. In other words, a complete set of normal and orthogonal functions is closed.

We will now prove that a closed set is complete. If it is not, there exists a function $f(x)$ belonging to L_2 such that

$$\lim_{n\to\infty} \int_a^b \left| f(x) - \sum_1^n \phi_k(x) \int_a^b f\bar{\phi}_k \, d\xi \right|^2 dx > 0, \tag{11.043}$$

for if equality held the set would be complete. From this and (11.04) we have

$$\int_a^b |f(x)|^2 \, dx - \sum_1^\infty \left| \int_a^b f\bar{\phi}_k \, dx \right|^2 > 0. \tag{11.044}$$

Let us set

$$g_n(x) = \sum_1^n \phi_k(x) \int_a^b f\bar{\phi}_k \, dx. \tag{11.05}$$

Then by the Riesz-Fischer theorem,

$$g(x) = \underset{n\to\infty}{\text{l.i.m.}}\; g_n(x) \tag{11.051}$$

exists, and

$$\int_a^b |g(x)|^2 \, dx = \lim_{n\to\infty} \int_a^b |g_n(x)|^2 \, dx = \sum_1^\infty \left| \int_a^b f\bar{\phi}_k \, dx \right|^2. \tag{11.052}$$

If we now consider $f(x) - g(x)$ we have

$$\int_a^b [f(x) - g(x)]\,\bar{\phi}_n(x)\,dx = \lim_{n\to\infty}\int_a^b [f(x) - g_n(x)]\,\overline{\phi_n(x)}\,dx = 0 \qquad (n = 1, 2, \cdots). \tag{11.053}$$

Since the set $\{\phi_n(x)\}$ is closed it follows that $f(x) - g(x)$ is equivalent to zero. But from (11.052) and (11.044),

$$\int_a^b |f|^2\,dx - \int_a^b |g|^2\,dx > 0, \tag{11.054}$$

and therefore $f(x)$ is not equivalent to $g(x)$. Thus a closed set is also complete. We have therefore proved that *a normal and orthogonal set of functions is closed when, and only when, it is complete.*

We now proceed to the general case where $\{f_n(x)\}$ is a denumerable set of functions all of which belong to L_2. We shall now construct a normal and orthogonal set from the given one. In the process of this construction we discard any term of $\{f_n(x)\}$ which is equivalent to zero or to a linear combination of terms preceding it. We now construct the sequence $\{\phi_k(x)\}$ where

$$\begin{aligned}
\phi_1(x) &= f_1(x)\Big/\left[\int_a^b |f_1(\xi)|^2\,d\xi\right]^{1/2};\\
\phi_2(x) &= \left[f_2(x) - \phi_1(x)\int_a^b f_2(\xi)\,\overline{\phi_1(\xi)}\,d\xi\right]\\
&\qquad\Big/\left[\int_a^b |f_2(\xi)|^2\,d\xi - \left|\int_a^b f_2\bar{\phi}_1\,d\xi\right|^2\right]^{1/2};\\
\phi_3(x) &= \left[f_3(x) - \phi_1(x)\int_a^b f_3(\xi)\,\overline{\phi_1(\xi)}\,d\xi - \phi_2(x)\int_a^b f_3(\xi)\,\overline{\phi_2(\xi)}\,d\xi\right]\\
&\qquad\Big/\left[\int_a^b |f_3|^2\,d\xi - \left|\int_a^b f_3\bar{\phi}_1\,d\xi\right|^2 - \left|\int_a^b f_3\bar{\phi}_2\,d\xi\right|^2\right]^{1/2};\\
&\cdots\cdots\cdots\cdots\cdots\cdots\cdots\cdots\cdots\cdots\cdots\cdots.
\end{aligned} \tag{11.06}$$

Since we discard all terms equivalent to zero or to a linear combination of preceding terms, none of the denominators will be zero. The set $\{\phi_k(x)\}$ is normal and orthogonal as can readily be verified. It is of the form

$$\begin{aligned}
\phi_1(x) &= b_{11} f_1(x);\\
\phi_2(x) &= b_{21} f_1(x) + b_{22} f_2(x);\\
\phi_3(x) &= b_{31} f_1(x) + b_{32} f_2(x) + b_{33} f_3(x);\\
&\cdots\cdots\cdots\cdots\cdots\cdots\cdots\cdots;
\end{aligned} \tag{11.061}$$

where no b_{nn} vanishes. From this representation it can be seen that the closure properties of the sets $\{\phi_k(x)\}$ and $\{f_k(x)\}$ are entirely equivalent. Further, since a polynomial in one set is a polynomial in the other set (for we can solve for $f_k(x)$ in terms of $\phi_k(x)$, since $b_{nn} \neq 0$) the completeness properties of the two sets are entirely equivalent. This allows us to extend the theorem proved for normal and orthogonal systems so that we now see that an enumerable sequence of functions is closed when, and only when, it is complete. This property of invariance of closure holds under very general transformations. We shall now show that closure is invariant under any linear transformation of the whole of L_2 into itself, which conserves the integral of the square of the modulus of each function. That is, *given a linear transformation such that to every function $f(x)$, of class L_2 over $a \leqq x \leqq b$,* corresponds $g(y)$, $c \leqq y \leqq d$ of class L_2, such that if $f_j(x) \rightarrow g_j(y)$,*

$$(11.07)\qquad \begin{gathered} cf_j(x) \rightarrow cg_j(y), \\ f_1(x) + f_2(x) \rightarrow g_1(y) + g_2(y), \\ \int_a^b |f_j(x)|^2\,dx = \int_c^d |g_j(y)|^2\,dy \qquad (j = 1, 2, \cdots). \end{gathered}$$

Then the closure properties of a sequence $\{f_n(x)\}$ are the same as those of $\{g_n(y)\}$.

For, by addition of

$$(11.071)\qquad \frac{1}{4}\int_a^b |f_1(x) + f_2(x)|^2\,dx = \frac{1}{4}\int_c^d |g_1(y) + g_2(y)|^2\,dy$$

and similar formulas concerning $f_1(x) - f_2(x)$, $f_1(x) + if_2(x)$, and $f_1(x) - if_2(x)$, with the respective factors $1, -1, i$, and $-i$, we obtain

$$(11.072)\qquad \int_a^b f_1(x)\,\overline{f_2(x)}\,dx = \int_c^d g_1(y)\,\overline{g_2(y)}\,dy.$$

Thus for a sequence $\{f_n(x)\}$ and a function $f(x)$ of L_2, with the corresponding g's as transforms, we have

$$(11.073)\qquad \int_a^b f(x)\,\overline{f_n(x)}\,dx = \int_c^d g(y)\,\overline{g_n(y)}\,dy \qquad (n = 1, 2, \cdots).$$

Thus the respective closure properties of the f_n and g_n sequences are equivalent.

Certain special cases of this theorem are of great importance. For example, if $\{f_n(\xi)\}$ is closed over the interval $(-\infty, \infty)$, so is the set of Fourier transforms

$$(11.08)\qquad g_n(\xi) = \underset{A\to\infty}{\text{l.i.m.}}\ \frac{1}{(2\pi)^{1/2}}\int_{-A}^{A} f_n(y)\,e^{iy\xi}\,dy.$$

For by the Plancherel and Parseval theorems this transformation falls under the class discussed above.

Again, let $\{f_n(x)\}$ be closed over (a, b). Let $\phi(t)$ be a function with a deriva-

* The sign $\rightarrow$ here means "corresponds to," not "tends to."

tive $\phi'(t)$ which is defined for every t and lies between positive finite limits for any range $(c - \epsilon, d + \epsilon)$. Let

$$\phi(c) = a; \qquad \phi(d) = b. \tag{11.09}$$

Then if $f(x)$ is any function belonging to L_2, and ψ is a real measurable function,

$$\int_a^b | f(x) |^2 dx = \int_c^d | f(\phi(t)) [\phi'(t)]^{1/2} e^{i\psi(t)} |^2 dt. \tag{11.10}$$

Thus the transformation which turns $f(x)$ into $f(\phi(t)) [\phi'(t)]^{1/2} e^{i\psi(t)}$ is reversible since $\phi'(t) > 0$, and leaves the class L_2 invariant, as well as the integral of the modulus of the square of every function. Like the Fourier transformation, it falls into the class discussed above and leaves the property of closure invariant.

An even simpler transformation leaving closure invariant changes $f(x)$ into $f(x)g(x)$, where $g(x)$ is a bounded measurable function independent of $f(x)$, and

$$\frac{1}{g(x)} > \epsilon > 0. \tag{11.11}$$

For if

$$\int_a^b f(x)h(x)\, dx = 0, \tag{11.12}$$

then

$$\int_a^b f(x)g(x) \cdot \frac{h(x)}{g(x)}\, dx = 0, \tag{11.13}$$

and vice versa. Clearly if $f(x)$ belongs to L_2, so do $f(x)g(x)$ and $f(x)/g(x)$. It follows at once that our transformation leaves closure invariant.

As a particular case, let us consider the set of functions $\{x^{\lambda_n}\}$ over the range $(0, 1)$. The well known Weierstrass approximation theorem tells us that if we take $\lambda_n = n \geqq 0$ then the system is closed. Here we shall study the more general case where we have given only that $\Re(\lambda_n) > -\frac{1}{2}$. This set is transformed into the set

$$\{e^{-(\lambda_n+1/2)\xi}\} \qquad (0 < \xi < \infty) \tag{11.14}$$

by the transformation

$$f(x) \rightarrow f(e^{-\xi})\, e^{-\xi/2}. \tag{11.15}$$

This again becomes

$$\{e^{-(\lambda_n+1/2)e^u}\, e^{u/2}\} \qquad (-\infty < u < \infty) \tag{11.16}$$

under the transformation

$$f(\xi) \rightarrow f(e^u)\, e^{u/2}. \tag{11.17}$$

Note that these transformations satisfy conditions of (11.07).

A Fourier transformation changes this set of functions into

$$\frac{1}{(2\pi)^{1/2}}\int_{-\infty}^{\infty} e^{-(\lambda_n+1/2)e^u}\, e^{u(1/2+iw)}\, du = \frac{1}{(2\pi)^{1/2}}\int_0^{\infty} e^{-(\lambda_n+1/2)v}\, v^{-1/2+iw}\, dv$$

$$(11.18)\qquad = \frac{1}{(2\pi)^{1/2}}(\lambda_n+\tfrac{1}{2})^{-1/2-iw}\,\Gamma(\tfrac{1}{2}+iw)$$

$$= \frac{1}{(2\pi)^{1/2}}(\lambda_n+\tfrac{1}{2})^{-1/2-iw}\,|\,\Gamma(\tfrac{1}{2}+iw\,|\,e^{i\psi(w)} \qquad (-\infty<w<\infty),$$

where $\psi(w)$ is a real function of w. By a well known theorem concerning the gamma function,

$$|\,\Gamma(\tfrac{1}{2}+iw)\,| = \{\Gamma(\tfrac{1}{2}+iw)\,\Gamma(\tfrac{1}{2}-iw)\}^{1/2}$$

$$= \{\pi \operatorname{sech} \pi w\}^{1/2}$$

$$(11.19)\qquad \sim (2\pi)^{1/2}\, e^{-\pi|w|/2}$$

$$\sim \left(\frac{\pi}{2}\right)^{1/2} \operatorname{sech}\frac{\pi w}{2}.$$

Thus using the various equivalence theorems developed above, the closure properties of $\{x^{\lambda_n}\}$ over $(0,1)$ are identical with those of $\{\pi(\operatorname{sech}\pi w/2)(\lambda_n+\frac{1}{2})^{-iw}\}$ and $\{e^{-\pi|w|/2}\,(\lambda_n+\frac{1}{2})^{-iw}\}$ over $(-\infty,\infty)$, for $\operatorname{sech}(\pi w/2) = g(w)\, e^{-\pi|w|/2}$ where $1 \leqq |\, g(w)\,| \leqq 2$, and therefore obeys the condition of (11.11).

By a second Fourier transformation, $\pi\,(\operatorname{sech}\pi w/2)\,(\lambda_n+\frac{1}{2})^{-iw}$ becomes

$$\left(\frac{\pi}{2}\right)^{1/2}\int_{-\infty}^{\infty} \operatorname{sech}\frac{\pi w}{2}\,(\lambda_n+\tfrac{1}{2})^{-iw}\, e^{-iwx}\, dw$$

$$= (2\pi)^{1/2}\int_0^{\infty}\frac{1}{e^{\pi w/2}+e^{-\pi w/2}}(\lambda_n+\tfrac{1}{2})^{-iw}\, e^{-iwx}\, dw$$

$$+ (2\pi)^{1/2}\int_{-\infty}^{0}\frac{1}{e^{\pi w/2}+e^{-\pi w/2}}(\lambda_n+\tfrac{1}{2})^{-iw}\, e^{-iwx}\, dw$$

$$= (2\pi)^{1/2}\int_0^{\infty}(e^{-\pi w/2}-e^{-3\pi w/2}+e^{-5\pi w/2}-\cdots)\,(\lambda_n+\tfrac{1}{2})^{-iw}\, e^{-iwx}\, dw$$

$$+ (2\pi)^{1/2}\int_{-\infty}^{0}(e^{\pi w/2}-e^{3\pi w/2}+e^{5\pi w/2}-\cdots)\,(\lambda_n+\tfrac{1}{2})^{-iw}\, e^{-iwx}\, dw$$

$$(11.20)\qquad = 2(2\pi)^{1/2}\sum_0^{\infty}\int_0^{\infty}(-1)^k\, e^{-(2k+1)\pi w/2}\cos(wx+w\log(\lambda_n+\tfrac{1}{2}))\, dw$$

$$= 2(2\pi)^{1/2}\sum_0^{\infty}(-1)^k\,\frac{\frac{2k+1}{2}\pi}{\left(\frac{2k+1}{2}\right)^2\pi^2+(x+\log(\lambda_n+\frac{1}{2}))^2}$$

$$= (2\pi)^{1/2} \operatorname{sech} (x + \log (\lambda_n + \tfrac{1}{2}))$$

$$= \frac{2(2\pi)^{1/2}}{(\lambda_n + \frac{1}{2}) e^x + \dfrac{1}{\lambda_n + \frac{1}{2}} e^{-x}} = \frac{2(2\pi)^{1/2} (\lambda_n + \frac{1}{2}) e^x}{(\lambda_n + \frac{1}{2})^2 e^{2x} + 1}.$$

We use the bounded convergence theorem above to invert summation and integration. The transformation

$$f(e^{2x})\, e^x \to f(x) \tag{11.21}$$

transforms this into

$$\frac{2(2\pi)^{1/2} (\lambda_n + \frac{1}{2})}{(\lambda_n + \frac{1}{2})^2 x + 1} \qquad (0 < x < \infty). \tag{11.22}$$

Thus the closure properties of

$$\frac{1}{(\lambda_n + \frac{1}{2})^2 x + 1}$$

over $(0, \infty)$ are the same as those of x^{λ_n} over $(0, 1)$.

12. **Szász's theorem.** Let us now investigate the closure of the set of functions (11.14). The quantities $\lambda_n + \frac{1}{2}$ all lie in the right half-plane. Thus the closure question is the same as that of the distribution of the zeros of the function

$$\phi(u) = \int_0^\infty F(\xi)\, e^{-u\xi}\, d\xi, \tag{12.01}$$

where $F(\xi)$ belongs to L_2 over $(0, \infty)$. For if $\phi(u)$ has an infinite number of zeros μ_n in the right half-plane, $\phi(\mu_n) = 0$, then $e^{-\mu_n \xi}$ cannot be a closed set. The function $\phi(u)$ will clearly be analytic in the right half-plane. Let $u = s + it$ where s and t are both real. Let us put

$$\psi(z) = \phi\left(\frac{1 - z}{1 + z}\right); \qquad \Phi(it) = \underset{A \to \infty}{\text{l.i.m.}} \int_0^A F(\xi)\, e^{-it\xi}\, d\xi;$$

(12.02)

$$\Psi (e^{i\theta}) = \Phi\left(\frac{1 - e^{i\theta}}{1 + e^{i\theta}}\right).$$

Then $\psi(z)$ is analytic in the unit circle. By exactly the same argument as in (8.13) $\psi(z)$ is the Poisson integral of $\Psi(e^{i\theta})$. Let the zeros of $\psi(z)$ in the circle $|z| < 1$, with those at the origin omitted and arranged so that their moduli will form a non-decreasing sequence, be $z_1, z_2, \cdots$. By (8.191), in view of Jensen's theorem, if $|z_n| \leqq r_n \leqq |z_{n+1}|$ and m is the smallest number such that $\psi^{(m)}(0) \neq 0$, we obtain

$$\log \left| \frac{r_n^n\, \psi^{(m)}(0)}{z_1 z_2 \cdots z_n} \right| = \frac{1}{2\pi} \int_{-\pi}^{\pi} \log |\psi(r_n e^{i\theta})|\, d\theta + \log m! - m \log r_n$$

$$(12.03)\qquad \leqq \frac{1}{2\pi}\int_{-\pi}^{\pi} \log^+ |\psi(r_n e^{i\theta})|\, d\theta + \text{const.}$$

$$\leqq \frac{1}{2\pi}\int_{-\pi}^{\pi} |\Psi(e^{i\theta})|\, d\theta + \text{const.};$$

and a fortiori, if $k < n$,

$$(12.04)\qquad \log\left|\frac{r_n^k\, \psi^{(m)}(0)}{z_1 z_2 \cdots z_k}\right| \leqq \frac{1}{2\pi}\int_{-\pi}^{\pi} |\Psi(e^{i\theta})|\, d\theta + \text{const.}$$

As a direct consequence of letting $r_n \to 1$ in (12.04) we get

$$(12.05)\qquad \log\left|\frac{1}{z_1 z_2 \cdots z_k}\right| < \text{const.},$$

and therefore

$$(12.051)\qquad \sum_1^{\infty} -\log |z_k| < \infty.$$

If we now invert the circle $|z| = 1$ back into $\mathfrak{J}(w) = 0$ and the zeros z_k of $\psi(z)$ into the zeros w_k of $\phi(w)$, this gives us

$$(12.06)\qquad w_k = \frac{1 - z_k}{1 + z_k}; \qquad z_k = \frac{1 - w_k}{1 + w_k};$$

and (12.05) becomes

$$(12.07)\qquad \infty > \sum_1^{\infty} \log\left|\frac{1 + w_k}{1 - w_k}\right| = \tfrac{1}{2}\sum_1^{\infty} \log\left[1 + \frac{2(w_k + \bar{w}_k)}{1 - w_k - \bar{w}_k + |w_k|^2}\right],$$

which is equivalent to

$$(12.071)\qquad \infty > \sum_1^{\infty} \frac{\mathfrak{R}(w_k)}{1 + |w_k|^2}.$$

On the other hand, let $\{z_k\}$ be a set of numbers satisfying

$$(12.08)\qquad \sum_1^{\infty} -\log |z_k| < \infty$$

and lying within the unit circle. From (12.08) we have

$$(12.09)\qquad \sum_1^{\infty} \frac{1 - |z_k|}{|z_k|} < \infty.$$

Within any circle $|z| < r < 1$ the product

$$(12.10)\qquad \prod_1^{\infty} \frac{z_k - z}{z_k - z\,|z_k|^2}$$

will converge uniformly to a function $h(z)$, which is analytic since

$$(12.11)\qquad \left|1 - \frac{z_k - z}{z_k - z\,|\,z_k\,|^2}\right| = \frac{|\,z\,|\,(1 - |z_k\,|^2)}{|\,z_k\,|\,|\,1 - z\,\bar{z}_k\,|} \leqq \frac{2}{1 - |\,z\,|}\,\frac{1 - |\,z_k\,|}{|\,z_k\,|},$$

and the majorant series converges by (12.09). Thus for any point inside the unit circle we have

$$(12.12)\qquad |\,h(z)\,| = \prod_1^\infty \left(\frac{1}{z_k}\right)^2 \prod_1^\infty \left|\frac{z - z_k}{z - \dfrac{1}{\bar{z}_k}}\right| \leqq \prod_1^\infty \left|\frac{1}{z_k}\right|^2 < \text{const.},$$

by (12.08). Thus $h(z)$ is an analytic function inside of the unit circle and is uniformly bounded in the entire circle. Let us put

$$(12.13)\qquad \psi(z) = h(z)\,(1 + z)^2.$$

Then $\psi(z)$ is defined in and on the unit circle and

$$(12.131)\qquad \psi(z_k) = 0 \qquad (k = 1, 2, \cdots).$$

Also $\psi(z)$ is uniformly bounded in the unit circle. Let us invert $|\,z\,| = 1$ into $\mathfrak{I}(w) = 0$ as before. We shall have

$$(12.14)\qquad \phi(w) = \psi\left(\frac{1 - w}{1 + w}\right).$$

From (12.12) and (12.13) we see that $\phi(s + it)$ is uniformly of class L_2 in t for any s such that $0 < s < \infty$. Accordingly, from Theorem V, we have

$$(12.15)\qquad \phi(w) = \int_0^\infty F(\xi)\, e^{-w\xi}\, d\xi$$

where $F(\xi)$ is of class L_2. Thus if

$$w_k = \frac{1 - 2\,k}{1 + 2\,k},$$

the set $\{e^{-w_k\xi}\}$ will not be closed over $(0, \infty)$ since $F(\xi)$ is of class L_2 and orthogonal to all of these by (12.15).

We have thus proved

THEOREM XIV. (SZÁSZ'S THEOREM.)* *Let λ_n be a set of numbers with real part exceeding $-1/2$. Then the set of functions $\{x^{\lambda_n}\}$ will be closed L_2 over $(0, 1)$ when, and only when,*

$$(12.16)\qquad \sum_{n=1}^\infty \frac{1 + 2\,\mathfrak{R}(\lambda_n)}{1 + |\,\lambda_n\,|^2} = \infty.$$

As a corollary we have:

* O. Szász, *Über die Approximation stetiger Funktionen durch lineare Aggregate von Potenzen*, Mathematische Annalen, vol. 77 (1916), pp. 482–496.

THEOREM XIV'.* *Let* σ_n *be a set of numbers in the strip* $|\Im(\sigma_n)| < \pi/2$. *Then the set of functions* $\{e^{\pi |w|/2 - \sigma_n i w}\}$ *will be closed over* $(-\infty, \infty)$ *when, and only when,*

$$\sum_{n=1}^{\infty} \frac{\cos \Im(\sigma_n)}{\cosh \Re(\sigma_n)} = \infty. \tag{12.17}$$

This follows at once from (11.19).

Again, by (11.22), we have

THEOREM XIV''.† *Let* $\{\mu_n\}$ *be a set of numbers with positive real part. Then the set of functions* $1/(\mu_n^2 x + 1)$ *will be closed over* $(0, \infty)$ *when, and only when*

$$\sum_{n=1}^{\infty} \frac{\Re(\mu_n)}{1 + |\mu_n|^2} = \infty. \tag{12.18}$$

Let us now turn from theorems of closure L_2 to theorems of the Weierstrass type, concerning the possibility of approximating uniformly to an arbitrary continuous function by polynomials in a given set of functions. Let

$$\int_0^1 \left| f(x)x^\alpha - \sum_1^n a_k x^{\lambda_k} \right|^2 dx < \epsilon. \tag{12.19}$$

Then by the Schwarz inequality,

$$\left| \int_0^x f(x)dx - \sum_1^n \frac{a_k \, x^{\lambda_k + 1 - \alpha}}{\lambda_k + 1 - \alpha} \right| \leqq \epsilon^{1/2} \left[\frac{1}{-2\alpha + 1} \right]^{1/2} \qquad (|\alpha| < \tfrac{1}{2}). \tag{12.20}$$

It is clearly always possible to approximate uniformly to an arbitrary continuous function vanishing at the origin by functions of the form

$$\int_0^x \phi(x) \, dx, \tag{12.21}$$

where $\phi(x)$ belongs to L_2.

On the other hand, let

$$\max_x \left| f(x) - \sum_1^n a_k \, x^{\lambda_k} \right| < \epsilon. \tag{12.22}$$

Then

$$\int_0^1 x^{2\alpha} \left| f(x) - \sum_1^n a_k \, x^{\lambda_k} \right|^2 dx < \frac{\epsilon^2}{2\alpha + 1} \qquad (\alpha > -\tfrac{1}{2}). \tag{12.23}$$

We have thus proved

* R. E. A. C. Paley and N. Wiener, *Notes on the theory and application of Fourier transforms; Note* IV, *a theorem on closure,* Transactions of the American Mathematical Society, vol. 35, pp. 766–768.

† This theorem has been proved by Professor Szász by different methods (unpublished paper). Cf. also Szász, *Über die Approximation stetiger Funktionen durch gegebene Funktionenfolgen,* Mathematische Annalen, vol. 104 (1931), pp. 155–160.

THEOREM XV. (SZÁSZ'S FORM OF MÜNTZ'S THEOREM.)* *Let* λ_n *be a set of numbers with positive real part. Then it will be possible to approximate uniformly over* (0, 1) *to an arbitrary continuous function by a polynomial* $C + \Sigma\, a_n x^{\lambda_n}$ *when*

$$\sum_{n=1}^{\infty} \frac{\Re(\lambda_n)}{1 + |\lambda_n|^2} = \infty, \tag{12.24}$$

and it will be impossible so to approximate for every continuous function when

$$\sum_{n=1}^{\infty} \frac{1 + \Re(\lambda_n)}{1 + |\lambda_n|^2} < \infty. \tag{12.25}$$

* Cf. Szász, loc. cit.; C. H. Müntz, *Über den Approximationssatz von Weierstrass*, Schwarz's Festschrift, Berlin, 1914, pp. 303–312.

CHAPTER III

CERTAIN INTEGRAL EXPANSIONS

13. **The integral equations of Laplace and of Planck.** The integral equation known by the name of Laplace is

$$g(u) = \int_0^\infty f(x)\, e^{-ux}\, dx, \tag{13.01}$$

where all arguments are real, $g(u)$ is given $(c < u < \infty)$ and $f(x)$ is sought. We shall assume the infinite integral to be taken in the sense $\lim_{A\to\infty} \int_0^A$ where $\int_0^A$ is a Lebesgue proper integral. Let us suppose that $g(a)$ exists and is finite, according to (13.01). Let us put

$$\int_0^x f(\xi)\, e^{-a\xi}\, d\xi = f_1(x). \tag{13.02}$$

We shall have for $u > a$,

$$g(u) = \int_0^\infty e^{(-u+a)x}\, df_1(x) = (u - a)x \int_0^\infty e^{(-u+a)x} f_1(x)\, dx. \tag{13.03}$$

Thus in case $u > a$, (13.01) will converge, and we shall have

$$\frac{g(a + \epsilon + u)}{\epsilon + u} = \int_0^\infty e^{-ux}\, [e^{-\epsilon x} f_1(x)]\, dx, \tag{13.04}$$

which we may write

$$g_2(u) = \int_0^\infty f_2(x)\, e^{-ux}\, dx. \tag{13.05}$$

The function $f_2(u)$ is by (13.02) bounded and is $O(e^{-\epsilon x})$ at infinity. Thus there is no essential restriction upon (13.01) if we assume that $f(x)$ belongs to L_2, or that it is bounded and continuous as well.

An exceedingly simple method of solving (13.01) is due to Widder.* By a direct differentiation, which we may readily justify,

$$(-1)^n\, g^{(n)}(u) = \int_0^\infty x^n f(x)\, e^{-ux}\, dx. \tag{13.06}$$

* D. V. Widder, *The inversion of the Laplace integral and the related moment problem*, Transactions of the American Mathematical Society, vol. 36, pp. 107–201.

Thus

$$\frac{(-1)^n g^{(n)}\left(\frac{n}{x}\right)\left(\frac{n}{x}\right)^{n+1}}{n!} = \frac{\int_0^\infty \xi^n f(\xi)\, e^{-n\xi/x}\, d\xi}{\int_0^\infty \xi^n e^{-n\xi/x}\, d\xi}. \tag{13.07}$$

Now let us consider

$$\psi_n(\xi) = \xi^n e^{-n\xi/x}. \tag{13.08}$$

This yields

$$\psi_n'(\xi) = \psi_n(\xi)\left[\frac{n}{\xi} - \frac{n}{x}\right]; \qquad \psi_n(\xi) \geqq 0 \qquad (0 < \xi < \infty). \tag{13.09}$$

Thus $\psi_n(\xi)$ has one maximum, and one only, when $x = \xi$. It is easy to show, moreover, that

$$\lim_{n\to\infty} \frac{\int_{x-\epsilon}^{x+\epsilon} \psi_n(\xi)\, d\xi}{\int^\infty \psi_n(\xi)\, d\xi} = \lim_{n\to\infty} \frac{\int_{1-\epsilon/x}^{1+\epsilon/x} \xi^n e^{-n\xi}\, d\xi}{\int_0^\infty \xi^n e^{-n\xi}\, d\xi} = 1. \tag{13.10}$$

That is, the expression of (13.07) represents an average of $f(\xi)$ with positive weighting, and as n becomes infinite, the total weight of the values of $f(\xi)$ for arguments outside $(x - \epsilon, x + \epsilon)$ tends to zero. Thus if $f(x)$ is bounded and continuous, a Fejér argument shows that

$$f(x) = \lim_{n\to\infty} \frac{(-1)^n g^{(n)}\left(\frac{n}{x}\right)\left(\frac{n}{x}\right)^{n+1}}{n!}. \tag{13.11}$$

Widder has shown how the restrictions of boundedness and continuity are not essential.

Another method of solving (13.01) is the following: if $f(x)$ belongs to L_2, so does $f(e^\xi)e^{\xi/2}$. Let us write

$$g(e^\eta)\, e^{\eta/2} = \int_{-\infty}^\infty f(e^\xi)\, e^{\xi/2}\, e^{(\xi+\eta)/2}\, e^{-e^{\xi+\eta}}\, d\xi. \tag{13.12}$$

Let us put

$$\begin{aligned} F(u) &= \underset{A\to\infty}{\mathrm{l.i.m.}} \frac{1}{(2\pi)^{1/2}} \int_{-A}^{A} f(e^\xi)\, e^{\xi(iu+1/2)}\, d\xi \\ &= \underset{\epsilon\to 0}{\mathrm{l.i.m.}} \frac{1}{(2\pi)^{1/2}} \int_\epsilon^{1/\epsilon} f(x)\, x^{iu-1/2}\, dx. \end{aligned} \tag{13.13}$$

By the Parseval theorem for Fourier transforms,

$$\frac{1}{(2\pi)^{1/2}} \int_{-\infty}^\infty g(e^\eta)\, e^{\eta/2}\, e^{iv\eta}\, d\eta = \int_{-\infty}^\infty e^{-e^\xi}\, e^{\xi(1/2+iv)}\, d\xi\, F(-v). \tag{13.14}$$

As we have already seen (cf. (11.18)),

$$\int_{-\infty}^{\infty} e^{-e^{\xi}} e^{\xi(1/2+iv)}\, d\xi = \Gamma(iv + \tfrac{1}{2}) = O\,(e^{-\pi\,|v|\,/2}). \tag{13.15}$$

Thus $\Gamma(\frac{1}{2} + iv)\, F(-v)$ belongs to L_2, and so do both its factors. By (1.7), we shall now have

$$\Gamma(\tfrac{1}{2} + iv)\, F(-\, v) = \underset{A\to\infty}{\text{l.i.m.}}\, \frac{1}{(2\pi)^{1/2}} \int_{-A}^{A} g(e^{\eta})\, e^{\eta/2}\, e^{iv\eta}\, d\eta. \tag{13.16}$$

Let it be noted that by Plancherel's theorem, $F(v)$ is a perfectly arbitrary function belonging to L_2, if $f(x)$ is itself an arbitrary function belonging to L_2. Thus if $g(u)$ is a function subject to the sole restriction that (13.01) can be solved by a function $f(x)$ belonging to L_2, we may replace this restriction by the equivalent restriction that

$$e^{\pm\pi v/2}\, \underset{A\to\infty}{\text{l.i.m.}} \int_{-A}^{A} g(e^{\eta})\, e^{\eta/2}\, e^{iv\eta}\, d\eta \tag{13.17}$$

both belong to L_2. A further appeal to Plancherel's theorem will formally turn this into the condition that

$$g(e^{\eta\pm\pi i/2})\, e^{\eta/2\pm\pi i/4} \tag{13.18}$$

both belong to L_2, or indeed that $g(\pm iy)$ both belong to L_2. This of course involves an extension of the definition of $g(u)$ to the complex domain. It will be seen that this formal result corresponds to the known result of Plancherel's theorem, that

$$\underset{A\to\infty}{\text{l.i.m.}} \int_{0}^{A} f(x)\, e^{\pm iyx}\, dx \tag{13.19}$$

belongs to L_2 over $(0, \infty)$.

By Plancherel's theorem, we may convert (13.16) into the form

$$\begin{aligned} f(e^{y})\, e^{y/2} &= \underset{A\to\infty}{\text{l.i.m.}}\, \frac{1}{(2\pi)^{1/2}} \int_{-A}^{A} e^{iy\xi}\, d\xi\, \frac{1}{\Gamma(i\xi + \frac{1}{2})} \\ &\qquad \times \underset{B\to\infty}{\text{l.i.m.}}\, \frac{1}{(2\pi)^{1/2}} \int_{-B}^{B} g(e^{w})\, e^{i\xi w + w/2}\, dw \\ &= \underset{A\to\infty}{\text{l.i.m.}}\, \frac{1}{2\pi} \int_{-\infty}^{\infty} g(e^{w})\, e^{w/2}\, dw \int_{-A}^{A} \frac{e^{i\xi(w+y)}\, d\xi}{\Gamma(i\xi + \frac{1}{2})}. \end{aligned} \tag{13.20}$$

We may write this in the form

$$f(x) = \underset{A\to\infty}{\text{l.i.m.}}\, \frac{1}{2\pi x} \int_{0}^{\infty} g\!\left(\frac{z}{x}\right) \frac{dz}{z^{1/2}} \int_{-A}^{A} \frac{z^{i\xi}\, d\xi}{\Gamma(i\xi + \frac{1}{2})}. \tag{13.21}$$

This yields us another solution of the Laplace equation.

We now turn to a physical problem involving an integral equation not unlike that of Laplace, which differs from it in that it leads us to the Riemann zeta function. By Planck's law the radiation per unit volume in a black cavity at temperature T in a state of equilibrium, and of frequency between ν and $\nu + d\nu$, is given by

$$\frac{8\pi\, h\nu^3}{c^3(e^{h\nu/(kT)} - 1)}\, d\nu, \tag{13.22}$$

where h is Planck's constant, c is the velocity of light, and k is the "gas constant," reckoned for one molecule. This suggests that the radiation from a source in approximate local equilibrium but consisting of a mixture of black bodies at different temperatures, will have a distribution given by

$$\nu^3\, d\nu \text{ const.} \int_0^\infty \frac{\phi(T)\, dT}{e^{h\nu/(kT)} - 1}, \tag{13.23}$$

where $\phi(T)$ represents in some sense the "amount" of radiation coming from black bodies at temperature T. If, then, we have an observable radiation with frequency distribution given by $\psi(\nu)d\nu$, the problem of resolving this into its constituent black-body radiations is equivalent to the solution of the equation

$$\psi(\nu) = \nu^3 \int_0^\infty \frac{\phi(T)\, dT}{e^{h\nu/(kT)} - 1}. \tag{13.24}$$

Let us now put

$$\frac{h}{kT} = \mu; \qquad \phi(T)\, dT = \mu\Phi(\mu)\, d\mu; \qquad \frac{\psi(\nu)}{\nu^2} = \Psi(\nu). \tag{13.25}$$

Then (13.24) assumes the form

$$\Psi(\nu) = \int_0^\infty \Phi(\mu)\, \frac{\mu\nu}{e^{\mu\nu} - 1}\, d\mu. \tag{13.26}$$

This we may write

$$\Psi(e^\eta)\, e^{\eta/2} = \int_{-\infty}^\infty \Phi(e^\xi)\, e^{\xi/2}\, \frac{e^{3(\xi+\eta)/2}\, d\xi}{e^{e^{\xi+\eta}} - 1}. \tag{13.27}$$

Let us put (as in (13.13))

$$F(u) = \underset{\epsilon\to\infty}{\text{l.i.m.}}\, \frac{1}{(2\pi)^{1/2}} \int_\epsilon^{1/\epsilon} \Phi(x)\, x^{iu-1/2}\, dx, \tag{13.28}$$

assuming that $\Phi(x)$ belongs to L_2. By the Parseval theorem for Fourier transforms,

$$\frac{1}{(2\pi)^{1/2}} \int_{-\infty}^\infty \Psi(e^\eta)\, e^{\eta/2}\, e^{iv\eta}\, d\eta = F(-v) \int_{-\infty}^\infty \frac{e^{\xi(3/2+iv)}\, d\xi}{e^{e^\xi} - 1}. \tag{13.29}$$

Now,

$$\int_{-\infty}^\infty \frac{e^{\xi(3/2+iv)}\, d\xi}{e^{e^\xi} - 1} = \int_0^\infty \frac{x^{1/2+iv}\, dx}{e^x - 1}$$

$$(13.30)\qquad = \Gamma\left(\frac{3}{2} + iv\right) \zeta\left(\frac{3}{2} + iv\right)$$

$$= O(v\, e^{-\pi |x| /2}).$$

Thus

$$\Gamma\left(\frac{3}{2} + iv\right) \zeta\left(\frac{3}{2} + iv\right)$$

and $F(-v)$ belong to L_2, as does their product. By (1.06) we shall have

$$(13.31)\qquad \Gamma\left(\frac{3}{2} + iv\right) \zeta\left(\frac{3}{2} + iv\right) F(-v) = \underset{A\to\infty}{\text{l.i.m.}}\ \frac{1}{(2\pi)^{1/2}} \int_{-A}^{A} \Psi(e^\eta)\, e^{\eta/2}\, e^{iv\eta}\, d\eta.$$

Thus we may determine $F(-v)$ and consequently $\Phi(x)$ and $\phi(T)$ in terms of $\Psi(x)$ and $\psi(\nu)$. This involves a division by the zeta function. In (13.31) $3/2 + iv$ is outside the critical strip, but it is easy to set up other formal solutions in which the division takes place within the critical strip, and in which the Riemann hypothesis on the zeros of the zeta function is consequently of moment.

14. **The integral equation of Stieltjes.** Let $g(u)$ be defined as in (13.01). Then formally

$$h(y) = \int_0^\infty e^{-uy}\, du \int_0^\infty f(x)\, e^{-ux}\, dx$$

$$(14.01)\qquad = \int_0^\infty f(x)\, dx \int_0^\infty e^{-u(x+y)}\, du$$

$$= \int_0^\infty \frac{f(x)\, dx}{x + y}.$$

This will now be proved using the theory of Fourier transforms. We have, for $y = e^\eta$,

$$(14.02)\qquad h(e^\eta)\, e^{\eta/2} = e^{\eta/2} \int_0^\infty \frac{f(x)\, dx}{x + y} = \int_{-\infty}^\infty \frac{f(e^\xi)\, e^{\xi/2}\, d\xi}{e^{(\xi-\eta)/2} + e^{(\eta-\xi)/2}}.$$

Now,

$$\frac{1}{(2\pi)^{1/2}} \int_{-\infty}^\infty \frac{e^{iy\xi}\, d\xi}{e^{\xi/2} + e^{-\xi/2}} = \left(\frac{2}{\pi}\right)^{1/2} \int_0^\infty \frac{\cos y\xi\, d\xi}{e^{\xi/2} + e^{-\xi/2}}$$

$$= \left(\frac{2}{\pi}\right)^{1/2} \int_0^\infty \cos y\xi\, (e^{-\xi/2} - e^{-3\xi/2} + e^{-5\xi/2} - \cdots)\, d\xi$$

$$(14.03)\qquad = \sum_{n=0}^\infty (-1)^n \left(\frac{2}{\pi}\right)^{1/2} \int_0^\infty \cos y\xi\, e^{-(2n+1)\xi/2}\, d\xi$$

$$= \sum_{n=0}^\infty (-1)^n \left(\frac{2}{\pi}\right)^{1/2} \frac{\dfrac{2n+1}{2}}{\left(\dfrac{2n+1}{2}\right)^2 + y^2} = \left(\frac{\pi}{2}\right)^{1/2} \operatorname{sech} \pi y.$$

Here we use the bounded convergence theorem to invert summation and integration. Thus by Parseval's theorem, if $f(x)$ belongs to L_2, if $F(y)$ is defined as in (13.13), and if we use (13.16),

$$
\begin{aligned}
\int_{-\infty}^{\infty} \frac{f(e^\xi)\, e^{\xi/2}\, d\xi}{e^{(\xi-\eta)/2} + e^{(\eta-\xi)/2}} &= \left(\frac{\pi}{2}\right)^{1/2} \int_{-\infty}^{\infty} F(y) \operatorname{sech} \pi y\, e^{-iy\eta}\, dy \\
&= \frac{1}{(2\pi)^{1/2}} \int_{-\infty}^{\infty} F(y)\, \Gamma(\tfrac{1}{2} + iy)\, \Gamma(\tfrac{1}{2} - iy)\, e^{-iy\eta}\, dy \\
&= \frac{1}{(2\pi)^{1/2}} \int_{\infty}^{\infty} \Gamma(\tfrac{1}{2} + iy)\, e^{-iy\eta}\, dy \\
&\qquad \times \operatorname*{l.i.m.}_{A\to\infty} \frac{1}{(2\pi)^{1/2}} \int_{-A}^{A} g(e^u)\, e^{u/2}\, e^{-iuy}\, du \\
&= \int_{-\infty}^{\infty} e^{-iy\eta}\, dy \frac{1}{(2\pi)^{1/2}} \int_{-\infty}^{\infty} e^{-e^\xi}\, e^{\xi/2}\, e^{iy\xi}\, d\xi \\
&\qquad \times \operatorname*{l.i.m.}_{A\to\infty} \frac{1}{(2\pi)^{1/2}} \int_{-A}^{A} g(e^u)\, e^{u/2}\, e^{-iuy}\, du \\
&= \int_{-\infty}^{\infty} e^{-e^{\xi+\eta}}\, e^{(\xi+\eta)/2}\, g(e^\xi)\, e^{\xi/2}\, d\xi \\
&= e^{\eta/2} \int_{\infty}^{\infty} g(x)\, e^{-xe^\eta}\, dx.
\end{aligned}
\tag{14.04}
$$

Combining this with (14.02), we see that

$$
h(y) = \int_0^{\infty} \frac{f(x)}{x+y}\, dy. \tag{14.01}
$$

Let us notice that the Maclaurin series for $\cosh y$ converges in such a way that its partial sums are all less in modulus than $\cosh y$. Let the $2m$th partial sum be

$$
\phi_m(y) = \sum_0^m \frac{(\pi y)^{2n}}{(2n)\,!}. \tag{14.05}
$$

Then

$$
\sum_0^m \frac{(-1)^n \left(\pi \dfrac{d}{d\eta}\right)^{2n}}{(2n)\,!} (h(e^\eta)\, e^{\eta/2}) = \left(\frac{\pi}{2}\right)^{1/2} \int_{-\infty}^{\infty} F(y) \operatorname{sech} \pi y\, \phi_m(y)\, e^{-iy\eta}\, dy. \tag{14.06}
$$

If $F(x)$ belongs to L_2 over $(-\infty, \infty)$,

$$
f(e^\xi)\, e^{\xi/2} = \operatorname*{l.i.m.}_{A\to\infty} \frac{1}{(2\pi)^{1/2}} \int_{-A}^{A} F(y)\, e^{-iy\eta}\, dy. \tag{14.07}
$$

Under the same circumstances,

$$\text{(14.08)}\qquad \begin{aligned} &\operatorname*{l.i.m.}_{A\to\infty} \left(\frac{1}{2\pi}\right)^{1/2} \int_{-A}^{A} F(y)\, e^{-iy\eta}\, dy \\ &\qquad = \operatorname*{l.i.m.}_{m\to\infty} \frac{1}{(2\pi)^{1/2}} \int_{-\infty}^{\infty} F(y) \operatorname{sech} \pi y\, \phi_m(y)\, e^{-iy\eta}\, dy, \end{aligned}$$

since by bounded convergence

$$\text{(14.09)}\qquad F(y) = \operatorname*{l.i.m.}_{m\to\infty} F(y) \operatorname{sech} \pi y\, \phi_m(y) = \lim_{m\to\infty} F(y) \operatorname{sech} \pi y\, \phi_m(y).$$

Combining (14.07) and (14.08), we get

$$\text{(14.10)}\qquad \pi f(e^{\xi})\, e^{\xi/2} = \operatorname*{l.i.m.}_{m\to\infty} \left(\frac{\pi}{2}\right)^{1/2} \int_{-\infty}^{\infty} F(y) \operatorname{sech} \pi y\, \phi_m(y)\, e^{-iy\eta}\, dy,$$

or by (14.06),

$$\text{(14.11)}\qquad f(e^{\xi})\, e^{\xi/2} = \operatorname*{l.i.m.}_{m\to\infty} \frac{1}{\pi} \sum_{0}^{m} (-1)^n \left(\pi \frac{d}{d\xi}\right)^{2n} [h(e^{\xi})\, e^{\xi/2}].$$

This we may write

$$\text{(14.12)}\qquad f(x) = \operatorname*{l.i.m.}_{m\to\infty} \frac{1}{\pi x^{1/2}} \sum_{0}^{m} (-1)^n \frac{\left(\pi x \dfrac{d}{dx}\right)^{2n}}{(2n)\,!} (x^{1/2}\, h(x)).$$

This gives us a simple solution of the Stieltjes integral equation. By reversing our steps, it is easy to show that the existence of the limit in the mean indicated in (14.11) is a necessary and sufficient condition for the existence of a solution to the Stieltjes equation. By (14.01) the Laplace equation will then have the solution

$$\text{(14.13)}\qquad \begin{aligned} f(x) &= \operatorname*{l.i.m.}_{m\to\infty} \frac{1}{\pi x^{1/2}} \sum_{0}^{m} (-1)^n \frac{\left(\pi x \dfrac{d}{dx}\right)^{2n}}{(2n)\,!} \left(x^{1/2} \int_0^{\infty} g(u)\, e^{-ux}\, du \right) \\ &= \operatorname*{l.i.m.}_{m\to\infty} \int_0^{\infty} g(u)\, K_m(ux)\, du, \end{aligned}$$

where

$$\text{(14.14)}\qquad K_m(w) = \frac{1}{\pi w^{1/2}} \sum_{0}^{m} (-1)^n \frac{\left(\pi w \dfrac{d}{dw}\right)^{2n}}{(2n)\,!} (w^{1/2}\, e^{-w}).$$

This will also yield another necessary and sufficient condition on $g(x)$ for the solvability of the Laplace equation in the case in which $f(x)$ belongs to L_2, to which we have shown all cases are reducible.

15. **An asymptotic series.** Formally* we have

$$h(y) = \int_0^\infty \frac{f(x)}{x+y}\,dx = \frac{1}{y}\int_0^\infty \frac{f(x)}{1+\frac{x}{y}}\,dx$$

(15.01)
$$= \frac{1}{y}\int_0^\infty f(x)\,dx - \frac{1}{y^2}\int_0^\infty xf(x)\,dx + \cdots$$
$$+ \frac{1}{y^{n+1}}\int_0^\infty (-x)^n f(x)\,dx + \int_0^\infty \frac{f(x)}{x+y}\left(\frac{x}{y}\right)^{n+1} dx.$$

Now if $x = e^\xi$, $y = e^\eta$, $f(e^\xi)e^{\xi/2} = \phi(\xi)$,

(15.02)
$$\int_0 \frac{f(x)}{x+y}\left(\frac{x}{y}\right)^{n+1} dx = \frac{e^{-\eta/2}}{2}\int_{-\infty}^\infty \phi(\xi)\,\text{sech}\,\frac{\xi-\eta}{2}\,e^{(\xi-\eta)\,(n+1)}\,d\xi,$$

and if $\phi(\xi)\,e^{(n+1)\xi}$ belongs to L_2,

(15.03)
$$\left|\int_0^\infty \frac{f(x)}{x+y}\left(\frac{x}{y}\right)^{n+1} dx\right| \leqq \frac{e^{-(n+3/2)\eta}}{2}$$
$$\times\left\{\int_{-\infty}^\infty |\,\phi(\xi)\,e^{(n+1)\xi}\,|^2\,d\xi \int_{-\infty}^\infty \text{sech}^2\,\frac{\xi-\eta}{2}\,d\xi\right\}^{1/2}$$
$$= O(y^{-(n+3/2)}).$$

Thus under this assumption, (15.01) yields the well known asymptotic series for $h(y)$. Let it be noted that the terms

(15.04)
$$\frac{1}{y}\int_0^\infty \left(\frac{x}{y}\right)^k f(x)\,dx = e^{-3\eta/2}\int_{-\infty}^\infty \phi(\xi)\,e^{(\xi-\eta)\,(k-1/2)}\,d\xi$$

have kernels $e^{-(\xi-\eta)\,(k-1/2)}$ whose formal Fourier transforms have singularities at the points $(k - \frac{1}{2})i$ which are also singularities of the Fourier transform $(\pi/2)^{1/2}$ sech πu of sech $((\xi + \eta)/2)$. This is not accidental, and in more general cases we shall find terms of an asymptotic series associated with the singularities of a kernel in the complex domain.

16. **Watson transforms.**† Let $\phi(x)$ be a function defined over $(-\infty, \infty)$, measurable, and with modulus everywhere 1. Let

* Cf. K. Knopp, *Theory and Application of Infinite Series*, London, 1928, last chapter.

† The theory of Watson transforms (G. N. Watson, *General transforms*, Proceedings of the London Mathematical Society, (2), vol. 35 (1932), pp. 156–199, has been recently subsumed under a more general theory of unitary transforms in Hilbert space by Bochner (*Inversion formulae and unitary transformations*, Annals of Mathematics, (2), vol. 34 (1934), pp. 111–115. The methods of this section are more akin to those of Hardy and Titchmarsh (*A class of Fourier kernels*, Proceedings of the London Mathematical Society, (2), vol. 35 (1932), pp. 116–155, and unlike those of Bochner, are specifically applicable to kernels of the form $K(xy)$, rather than to the more general kernels of the form $K(x, y)$. Cf. also Titchmarsh, Journal of the London Mathematical Society, vol. 8 (1933), pp. 217–220, Plancherel, ibid., vol. 8 (1933), pp. 220–226, and Ida W. Busbridge, ibid., vol. 9 (1934), pp. 179–186.

$$\phi_a(x) = \frac{2}{\pi a} \int_{-\infty}^{\infty} \phi(x - \xi) \frac{\sin^2 \frac{a\xi}{2}}{\xi^2} d\xi. \tag{16.01}$$

belong to L_2 for every a. Let

$$\Phi_a(u) = \underset{A\to\infty}{\text{l.i.m.}} \frac{1}{(2\pi)^{1/2}} \int_{-A}^{A} \phi_a(x)\, e^{iux}\, dx. \tag{16.02}$$

Then if $a < b < c$,

$$\begin{aligned}
&\int_{-a}^{a} \left[1 - \frac{|u|}{a}\right]^2 |\Phi_b(u) - \Phi_c(u)|^2\, du \\
&= \int_{-\infty}^{\infty} dx \left| \frac{1}{(2\pi)^{1/2}} \int_{-a}^{a} \left[1 - \frac{|u|}{a}\right] \Phi_b(u)\, e^{-iux}\, du \right. \\
&\qquad \left. - \frac{1}{(2\pi)^{1/2}} \int_{-a}^{a} \left[1 - \frac{|u|}{a}\right] \Phi_c(u)\, e^{-iux}\, du \right|^2 \\
&= \int_{-\infty}^{\infty} dx \left| \frac{2}{\pi a} \int_{-\infty}^{\infty} \frac{\sin^2 \frac{a(x-\xi)}{2}}{(x-\xi)^2} d\xi \right. \\
&\qquad \left. \times \frac{2}{\pi} \int_{-\infty}^{\infty} \left(\frac{\sin^2 \frac{b\eta}{2}}{b\eta^2} - \frac{\sin^2 \frac{c\eta}{2}}{c\eta^2} \right) \phi(\xi - \eta)\, d\eta \right|^2 \\
&= \int_{-\infty}^{\infty} dx \left| \frac{4}{\pi^2 a} \int_{-\infty}^{\infty} \left(\frac{\sin^2 \frac{b\eta}{2}}{b\eta^2} - \frac{\sin^2 \frac{c\eta}{2}}{c\eta^2} \right) d\eta \right. \\
&\qquad \left. \times \int_{-\infty}^{\infty} \phi(\xi) \frac{\sin^2 a(x - \eta - \xi)}{(x - \eta - \xi)^2} d\xi \right|^2 \\
&= \int_{-a}^{a} \left| \Phi_a(u) \frac{|u|(b - c)}{bc} \right|^2 du.
\end{aligned} \tag{16.03}$$

It immediately results that

$$\lim_{b,c\to\infty} \int_{-a}^{a} \left[1 - \frac{|u|}{a}\right]^2 |\Phi_b(u) - \Phi_c(u)|^2\, du = 0, \tag{16.04}$$

and hence there exists over every finite range a function

$$\Phi(u) = \underset{b\to\infty}{\text{l.i.m.}}\, \Phi_b(u). \tag{16.05}$$

Let us now start with a function $f(x)$ such that

$$f(x) = 0 \qquad [\,|x| > A], \tag{16.06}$$

and let us put

$$(16.07)\quad g(y) = \frac{1}{(2\pi)^{1/2}}\int_{-\infty}^{\infty} f(x)\,\Phi(y - x)\,dx = \lim_{b\to\infty} \frac{1}{(2\pi)^{1/2}}\int_{-\infty}^{\infty} f(x)\,\Phi_b(y - x)\,dx;$$

$$(16.08)\qquad g_b(y) = \frac{1}{(2\pi)^{1/2}}\int_{-\infty}^{\infty} f(x)\,\Phi_b(y - x)\,dx.$$

We shall have

$$(16.09)\qquad \frac{1}{(2\pi)^{1/2}}\int_{-\infty}^{\infty} e^{-iuy}\,dy\,\frac{1}{(2\pi)^{1/2}}\int_{-\infty}^{\infty} f(x)\,\Phi_b(y - x)\,dx = \frac{\phi_b(u)}{(2\pi)^{1/2}}\int_{-\infty}^{\infty} f(x)\,e^{-iux}\,dx,$$

and hence

$$(16.10)\qquad \int_{-\infty}^{\infty} |\,g_b(y)\,|^2\,dy = \int_{-\infty}^{\infty} |\,\phi_b(u)\,|^2 \left|\frac{1}{(2\pi)^{1/2}}\int_{-\infty}^{\infty} f(x)\,e^{-iux}\,dx\right|^2 du.$$

By (16.01) and Fejér's theorem, we have almost everywhere

$$(16.11)\qquad |\,\phi_b(u)\,| \leqq 1; \qquad \lim_{b\to\infty} |\,\phi_b(u)\,| = 1;$$

so that by (16.10)

$$(16.12)\qquad \int_{-\infty}^{\infty} |\,g_b(y)\,|^2\,dy \leqq \int_{-\infty}^{\infty} \left|\frac{1}{(2\pi)^{1/2}}\int_{-\infty}^{\infty} f(x)\,e^{-iux}\,dx\right|^2 du = \int_{-\infty}^{\infty} |\,f(x)\,|^2\,dx,$$

and by bounded convergence,

$$(16.13)\qquad \lim_{b\to\infty}\int_{-\infty}^{\infty} |\,g_b(y)\,|^2\,dy = \int_{-\infty}^{\infty} |\,f(x)\,|^2\,dx.$$

Returning to (16.09), we have

$$(16.14)\qquad \lim_{b,c\to\infty}\int_{-\infty}^{\infty} |\,g_b(y) - g_c(y)\,|^2\,dy = \lim_{b,c\to\infty}\int_{-\infty}^{\infty} |\,\phi_b(u) - \phi_c(u)\,|^2 \times \left|\frac{1}{(2\pi)^{1/2}}\int_{-\infty}^{\infty} f(x)\,e^{-iux}\,dx\right|^2 du = 0.$$

From this, (16.07), and the often used principle that where a limit and a limit in the mean both exist, they are equivalent,

$$(16.15)\qquad \underset{b\to\infty}{\text{l.i.m.}}\; g_b(y) = g(y).$$

Thus by (16.13) we get

$$\int_{-\infty}^{\infty} | g(y) |^2 \, dy = \int_{-\infty}^{\infty} | f(x) |^2 \, dx. \tag{16.16}$$

Let us now take up the case in which $f(x)$ is a perfectly general function belonging to L_2. We have by (16.16), applied to the function which is f over (A, B) and $(-B, -A)$ and zero elsewhere,

$$\begin{aligned} \int_{-\infty}^{\infty} \left| \frac{1}{(2\pi)^{1/2}} \int_{-B}^{B} f(x) \, \Phi(y - x) \, dx - \frac{1}{(2\pi)^{1/2}} \int_{-A}^{A} f(x) \, \Phi(y - x) \, dx \right|^2 dy \\ = \int_{-B}^{B} | f(x) |^2 \, dx - \int_{-A}^{A} | f(x) |^2 \, dx. \end{aligned} \tag{16.17}$$

Thus by the Riesz-Fischer theorem, there exists a function

$$g(y) = \underset{A\to\infty}{\text{l.i.m.}} \frac{1}{(2\pi)^{1/2}} \int_{-A}^{A} f(x) \, \Phi(y - x) \, dx \tag{16.18}$$

belonging to L_2, and we have

$$\int_{-\infty}^{\infty} | g(y) |^2 \, dy = \int_{-\infty}^{\infty} | f(x) |^2 \, dx. \tag{16.19}$$

Now by (16.08), (16.15), and (16.09), if $f(x)$ vanishes for sufficiently large arguments,

$$\begin{aligned} \underset{A\to\infty}{\text{l.i.m.}} \frac{1}{(2\pi)^{1/2}} \int_{-A}^{A} e^{-iuy} \, dy \frac{1}{(2\pi)^{1/2}} \int_{-\infty}^{\infty} f(x) \, \Phi(y - x) \, dx \\ = \frac{\phi(u)}{(2\pi)^{1/2}} \int_{-\infty}^{\infty} f(x) \, e^{-iux} \, dx. \end{aligned} \tag{16.20}$$

By (16.18), if we approximate to $f(x)$ by functions differing from 0 only over a finite region, and then proceed to the limit, we may extend (16.20) to the general case where $f(x)$ belongs to L_2. This gives us

$$\underset{A\to\infty}{\text{l.i.m.}} \frac{1}{(2\pi)^{1/2}} \int_{-A}^{A} g(y) \, e^{-iuy} \, dy = \underset{B\to\infty}{\text{l.i.m.}} \frac{1}{(2\pi)^{1/2}} \int_{-B}^{B} f(x) \, e^{-iux} \, dx \, \phi(u). \tag{16.21}$$

Let us now remember that $\overline{\phi(u)} = 1/\phi(u)$. We obtain

$$\underset{B\to\infty}{\text{l.i.m.}} \frac{1}{(2\pi)^{1/2}} \int_{-B}^{B} f(x) \, e^{-iux} \, dx = \overline{\phi(u)} \, \underset{A\to\infty}{\text{l.i.m.}} \frac{1}{(2\pi)^{1/2}} \int_{-A}^{A} g(y) \, e^{-iuy} \, dy, \tag{16.22}$$

or by a further Fourier transformation, as in (16.20),

$$f(x) = \underset{A\to\infty}{\text{l.i.m.}} \frac{1}{(2\pi)^{1/2}} \int_{-A}^{A} g(y) \, \overline{\Phi}(y - x) \, dy. \tag{16.23}$$

If we now put

$$f(-x) = h(x), \tag{16.24}$$

then if $h(x)$ belongs to L_2, we obtain the dual pair of formulas

$$\text{(16.25)}\qquad \begin{aligned} g(y) &= \operatorname*{l.i.m.}_{A\to\infty} \frac{1}{(2\pi)^{1/2}} \int_{-A}^{A} h(x)\, \Phi(x+y)\, dx; \\ h(x) &= \operatorname*{l.i.m.}_{A\to\infty} \frac{1}{(2\pi)^{1/2}} \int_{-A}^{A} g(y)\, \overline{\Phi}(x+y)\, dy. \end{aligned}$$

Let us now make the transformations

$$\text{(16.26)}\qquad \begin{gathered} y = \log \eta; \qquad x = \log \xi; \qquad g(y) = \eta^{1/2}\, G(\eta); \qquad h(x) = \xi^{1/2}\, H(\xi); \\ \Phi(x) = \xi^{1/2}\, K(\xi). \end{gathered}$$

Then (16.25) becomes

$$\text{(16.27)}\qquad \begin{aligned} G(\eta) &= \operatorname*{l.i.m.}_{\epsilon\to 0} \frac{1}{(2\pi)^{1/2}} \int_{\epsilon}^{1/\epsilon} H(\xi)\, K(\xi\eta)\, d\xi; \\ H(\xi) &= \operatorname*{l.i.m.}_{\epsilon\to 0} \frac{1}{(2\pi)^{1/2}} \int_{\epsilon}^{1/\epsilon} G(\eta)\, \overline{K(\xi\eta)}\, d\eta. \end{aligned}$$

By (16.19),

$$\text{(16.28)}\qquad \int_0^\infty |\, G(\eta)\, |^2\, d\eta = \int_0^\infty |\, H(\xi)\, |^2\, d\xi.$$

We append a table of values of $\phi(x)$ and corresponding values of $K(\xi)$.

$\phi(x)$	$K(\xi)$
$\left(\frac{2}{\pi}\right)^{1/2} \Gamma(\frac{1}{2} - ix) \cos \frac{\pi}{2} (\frac{1}{2} - ix).$	$2 \sin \xi$ (sine transform).
$-i\left(\frac{2}{\pi}\right)^{1/2} \Gamma(\frac{1}{2} - ix) \sin \frac{\pi}{2} (\frac{1}{2} - ix).$	$2 \cos \xi$ (cosine transform).
$e^{ie^x}.$	$\dfrac{\xi^{(\pi-1)/2}\, i\, \Gamma(i \log \xi)}{2\pi}.$
$e^{ix^2/2}.$	$\dfrac{1-i}{2}\, \xi^{-1/2-i\xi/2}.$
$\dfrac{2^{ix}\, \Gamma\left(\frac{m+ix}{2} + \frac{1}{2}\right)}{\Gamma\left(\frac{m-ix}{2} + \frac{1}{2}\right)}.$	$(2\pi)^{1/2} J_m(\xi)\, \xi^{1/2}$ (Hankel transform).

Chapter IV

A class of singular integral equations

17. The theory of Hopf and Wiener.† We shall devote this section to the solution of the homogeneous linear integral equation

$$f(x) = \int_0^\infty K(x - y) f(y) \, dy, \tag{17.01}$$

where we assume that the kernel K vanishes exponentially for large values of $|x|$. The simplest particular case is the equation of Lalesco,

$$K(x) = e^{-|x|}; \tag{17.02}$$

Milne's equation is

$$K(x) = \tfrac{1}{2} \int_{|x|}^\infty \frac{e^{-t}}{t} \, dt. \tag{17.03}$$

This equation yields as its solution the distribution of temperature in a stellar atmosphere in radiative equilibrium.‡ We shall apply the method of Fourier transforms together with certain elementary considerations from the theory of functions of a complex variable to obtain the "fundamental solutions" of (17.01): that is, all solutions $f(x)$ which become infinite for great values of x more slowly than an exponential function with a smaller exponent than the reciprocal of one dominating the kernel $K(x)$. We shall represent the solutions in explicit integral form.

The analogous equation

$$f(x) = \int_{-\infty}^\infty K(x - y) f(y) \, dy \tag{17.04}$$

has a much simpler theory. Its solutions are essentially exponential functions. If $u = u^*$ is an n-fold zero of the function

$$1 - \int_{-\infty}^\infty K(t) \, e^{ut} \, dt, \tag{17.05}$$

if Q is an arbitrary polynomial of degree not exceeding $n - 1$, and if all the necessary integrals have a sense,

$$Q(x) \, e^{-u^* x} \tag{17.06}$$

† N. Wiener and E. Hopf, *Über eine Klasse singulärer Integralgleichungen*, Sitzungsberichte der Preussischen Akademie, Mathematisch-Physikalische Klasse, 1931, p. 696.

‡ E. Hopf, *Mathematisches zur Strahlungsgleichgewichtstheorie der Fixsternatmosphären*, Mathematische Zeitschrift, vol. 33 (1931), p. 109.

is a solution of (17.04). We may express the relation of (17.01) to (17.04) as follows: the solution of (17.01) behaves asymptotically for large x in the same way as certain solutions of (17.04). This fact is already to be seen in the two examples mentioned above.

We now proceed to the solution of (17.01). Let the kernel K be real and continuous except for a finite number of finite jumps. Let $K(x)\, e^{s|x|}$ belong to L_2 for at least one positive s. There is no real restriction in assuming that the

$$(17.07) \qquad \int_{-\infty}^{\infty} (K(x)\, e^{s|x|})^2\, dx$$

converge for all $s < 1$. Then, by the Schwarz inequality,

$$(17.08) \qquad \int_{-\infty}^{\infty} |\, K(x)\,|\, e^{s|x|}\, dx$$

converges for $s < 1$, as we may see by writing the integrand in the form

$$K(x)\, e^{(s+1)|x|/2}\, e^{-(1-s)|x|/2}.$$

In what follows, we consider those solutions of (17.01) for which

$$(17.09) \qquad f(x) = O(e^{\alpha x}),$$

where α is an arbitrary fixed constant less than 1.

In order to solve (17.01) under these assumptions, we first write it in the form

$$(17.10) \qquad g(x) = f(x) - \int_{-\infty}^{\infty} K(x-y)\, f(y)\, dy,$$

where

$$(17.11) \qquad f(x) = 0 \quad (x < 0); \qquad g(x) = 0 \quad (x > 0).$$

Here $g(x)$ is defined for $x < 0$ by the right side of (17.10). We now introduce the Laplace transforms

$$(17.12) \quad \phi(u) = \int_{-\infty}^{\infty} f(x)\, e^{ux}\, dx, \quad \gamma(u) = \int_{-\infty}^{\infty} g(x) e^{ux}\, dx, \quad \kappa(u) = \int_{-\infty}^{\infty} K(x)\, e^{ux}\, dx$$

where $u = s + it$ is a complex variable. Then

$$\frac{\phi(s+it)}{(2\pi)^{1/2}}, \qquad \frac{\gamma(s+it)}{(2\pi)^{1/2}}, \qquad \frac{\kappa(s+it)}{(2\pi)^{1/2}},$$

considered as functions of t, are the Fourier transforms of $f(x)\, e^{sx}$, $g(x)\, e^{sx}$, $K(x)\, e^{sx}$. By (17.11),

$$(17.13) \qquad \phi(u) = \int_0^{\infty} f(x)\, e^{ux}\, dx,$$

$$(17.14) \qquad \gamma(u) = \int_{-\infty}^{0} g(x)\, e^{ux}\, dx.$$

By Theorems V and III, $\phi(u)$ is regular in the half-plane $s < -\alpha$, and is bounded in every properly included half-plane. The same reference shows that $\gamma(u)$ is regular in the half-plane $s > -1$ and is bounded in every properly included half-plane since $g(x) = O(e^{-|x|})$.

Now, the half-planes of regularity of ϕ and of γ together exhaust the entire u-plane and have a strip $(-1 < s < -\alpha)$ in common. Under our assumptions concerning $K(x)$, the Laplace integral of $K(x)$ converges absolutely for $|s| < 1$, and $K(u)$ is regular in this strip. If we now apply the Laplace transformation to (17.10), we get

$$\begin{aligned}\gamma(u) &= \phi(u) - \int_{-\infty}^{\infty} e^{ux}\,dx \int_{-\infty}^{\infty} K(x-y)\,f(y)\,dy \\ &= \phi(u) - \int_{-\infty}^{\infty} f(y)\,dy \int_{-\infty}^{\infty} K(x-y)\,e^{ux}\,dx \qquad (17.15)\\ &= \phi(u) - \int_{-\infty}^{\infty} f(y)\,e^{uy}\,dy \int_{-\infty}^{\infty} K(t)\,e^{ut}\,dt;\end{aligned}$$

or in other words,

$$\gamma(u) = \phi(u)\,(1 - \kappa(u)). \tag{17.16}$$

Since the integrals converge absolutely, in case $-1 < s = \Re(u) < -\alpha$, the indicated inversion of the order of integration is legitimate.

In order to apply (17.16), we need the following:

Lemma. *The function $1 - \kappa(u)$ possesses at most a finite set of zeros in every strip $|\sigma| \leqq \beta$ $(\beta < 1)$. If we represent these by $u_1, u_2, \cdots, u_m$, then we may represent $1 - \kappa(u)$ in the strip $|\sigma| \leqq \beta$ in the form*

$$1 - \kappa(u) = \frac{\sigma_+(u)}{\sigma_-(u)} \prod_1^m (u - u_\nu). \tag{17.17}$$

Here $\sigma_+(u)$ is regular and free from zeros in the half-plane $s \geqq -\beta$, while $\sigma_-(u)$ is regular and free from zeros in the half-plane $s \leqq +\beta$. The moduli

$$|\sigma_+(u)\,u^{k+m/2}|, \qquad |\sigma_-(u)\,u^{k-m/2}|, \tag{17.18}$$

lie for sufficiently large u between positive bounds in their respective half-planes. Here k is a definite integer, determined by $K(x)$. For an even kernel, $K = K(|x|)$, we have $\kappa(u) = \kappa(-u)$, m is an even number $2n$, and $k = 0$.

If we put $x = \xi + \pi/t$ we see that

$$\kappa(u) = \int_{-\infty}^{\infty} K(x)\,e^{sx}\,e^{itx}\,dx = -\int_{-\infty}^{\infty} K\left(\xi + \frac{\pi}{t}\right) e^{s(\xi+\pi/t)}\,e^{it\xi}\,d\xi. \tag{17.19}$$

If we now replace ξ in the second integral by x, and add the second integral to the first, we get

$$2\kappa(u) = \int_{-\infty}^{\infty} \left\{K(x)\,e^{sx} - K\left(x + \frac{\pi}{t}\right) e^{s(x+\pi/t)}\right\} e^{itx}\,dx. \tag{17.20}$$

By splitting this into two parts, we may see that the modulus of the integrand does not exceed

(17.21) $$e^{\pi/|t|}\left|K\left(x+\frac{\pi}{t}\right)-K(x)\right|e^{sx}+|e^{s\pi/t}-1|\,|K(x)|\,e^{sx}$$

for $|s|<1$. Hence if $|s|\leqq s_0<1$, it follows that

(17.22) $$\begin{aligned}2|\kappa(u)|\leqq e^{\pi/|t|}\int_{-\infty}^{\infty}\left|K\left(x+\frac{\pi}{t}\right)-K(x)\right|e^{s_0|x|}\,dx+|e^{s\pi/t}-1|\\ \times\int_{-\infty}^{\infty}|K(x)|\,e^{s_0|x|}\,dx.\end{aligned}$$

If we now make use of our assumptions concerning K, it readily follows that

(17.23) $$\kappa(s+it)\to 0 \quad\text{for}\quad |t|\to\infty$$

uniformly in the strip $|s|\leqq s_0$. From this our assertion concerning the zeros of $1-\kappa(u)$ follows at once. Moreover,

(17.24) $$\frac{\kappa(s+it)}{(2\pi)^{1/2}}$$

taken as a function of t, is the Fourier transform of $K(x)\,e^{sx}$, which we have already assumed to belong to L_2. Thus by Plancherel's theorem, $|\kappa(s+it)|$ also belongs to L_2 for $-\infty<t<+\infty$.

Let us now put

(17.25) $$\tau(u)=(1-\kappa(u))\frac{(u^2-1)^{m/2}}{\prod_1^m(u-u_\nu)}\left(\frac{u+1}{u-1}\right)^k,$$

where k is still to be determined, and where we understand by $(u^2-1)^{m/2}$ that single-valued branch in $|s|<1$ which behaves like u^m for large values of $|u|$. Let $\beta<\beta'<1$, where however we so choose β' that the somewhat larger strip $|s|\leqq\beta'$ contains no new zeros of $1-\kappa(u)$. Then $\tau(u)$ is regular and free from zeros in $|s|\leqq\beta'$, and we have

(17.26) $$\tau(u)\to 1 \quad\text{for}\quad |t|\to\infty,$$

uniformly over $|s|\leqq\beta'$.

Now let us consider the increase of $\log\tau(s+it)$, as t runs from $-\infty$ to $+\infty$. By (17.26), it is an integral multiple of $2\pi i$, and can be made to vanish by a suitable choice of k in (17.25). After we have so determined k, let us consider that branch of $\log\tau(u)$ which is single-valued over the strip $|s|\leqq\beta'$, for which $\log\tau(u)$ tends to 0 as $t\to-\infty$. We shall also have $\log\tau(u)\to 0$ as $t\to+\infty$, or

(17.27) $$\log\tau(u)\to 0 \quad\text{for}\quad |t|\to\infty$$

throughout the strip $|s|\leqq\beta'$. Moreover, $\tau(u)$ is of the form

(17.28) $$(1-\kappa(u))\left(1+O\left(\frac{1}{|u|}\right)\right)$$

for large $| t |$ so that $| \log \tau(u) |$ also belongs to L_2 in t. We may then apply the Cauchy integral formula and we get $\log \tau = \log \tau_+ - \log \tau_-$ where

$$\log \tau_+(u) = -\frac{1}{2\pi i}\int_{-\beta'-i\infty}^{-\beta'+i\infty} \frac{\log \tau(v)}{v-u}\,dv,$$

$$(17.29)\qquad \log \tau_-(u) = -\frac{1}{2\pi i}\int_{\beta'-i\infty}^{\beta'+i\infty} \frac{\log \tau(v)}{v-u}\,dv,$$

for $-\beta' < s = \Re(u) < \beta'$.

Now, by the Schwarz inequality,

$$(17.30)\qquad | \log \tau_-(u) |^2 \leqq \frac{1}{4\pi^2}\int_{\beta'-i\infty}^{\beta'+i\infty} | \log \tau(v) |^2 \, | dv | \int_{\beta'-i\infty}^{\beta'+i\infty} \frac{| dv |}{| u - v |^2},$$

so that $\log \tau_-(u)$ is regular and bounded for $s \leqq \beta < \beta'$. Similarly, $\log \tau_+(u)$ is regular and bounded for $s \geqq -\beta > -\beta'$. If we now put

$$(17.31)\qquad \sigma_+(u) = \tau_+(u)\,(u+1)^{-k-m/2}, \qquad \sigma_-(u) = \tau_-(u)\,(u-1)^{-k+m/2},$$

then (17.17) and all the properties we have claimed for σ_+ and σ_- follow from (17.25) and (17.29).

If K is even, clearly $\kappa(u) = \kappa(-u)$ and the number of roots of $1 - \kappa(u)$ in the strip $| s | \leqq \beta$ is even, $m = 2n$. Since $K(x)$ is real, $\overline{\kappa(u)} = \kappa(\bar{u})$. If now u is imaginary, $\bar{u} = -u$ and $\overline{\kappa(u)} = \kappa(-u) = \kappa(u)$. That is, $\kappa(u)$ is real. If we put $k = 0$ in (17.25), $\tau(u)$ is also real for imaginary u. Since $\tau(u) \neq 0$, the increment of $\log \tau(u)$ along the imaginary u-axis is zero. It should be emphasized that the functions σ_+ and σ_- are determined by the kernel $K(x)$ through explicit integral formulas.

It is now easy to solve (17.16). Let $u_1, u_2, \cdots, u_m$ be all the zeros of $1 - \kappa(u)$ lying in the strip $| s | \leqq \alpha$, α being taken as in (17.09). Let us determine β so that $\alpha < \beta < 1$, while no new zeros lie in the strip $| s | \leqq \beta$. Using this β, we apply our lemma. Then (17.16) and (17.17) yield

$$(17.32)\qquad \frac{\gamma(u)}{\sigma_+(u)} = \frac{\phi(u)}{\sigma_-(u)} \prod_1^m (u - u_\nu).$$

By our lemma and another remark we have already made, the left side of this equation is regular for $s \geqq -\beta$, and is of the form $O(| u |^{k+m/2})$ for large $| u |$. Similarly, the right side is regular for $s \leqq -\beta$ and is of the form $O(| u |^{k+m/2})$ for large $| u |$. Thus (17.32) defines an entire function which is $O(| u |^{k+m/2})$ for large $| u |$, and can only be a polynomial of degree not exceeding $k + m/2$.

In what follows, we confine ourselves to the case of an even real kernel $K(x) = K(| x |)$. Then $m = 2n$, $k = 0$, and (17.32) defines a polynomial of at most the nth degree. However, the nth degree itself is excluded, for otherwise, by our lemma, $| \phi(u) |$ will lie above a positive bound for large $| u |$, which contradicts the fact that by Plancherel's theorem, $|\phi(s + it)|$ belongs to L_2 as a function of t. Thus

$$\phi(u) = \frac{\sigma_-(u)\, P_{n-1}(u)}{\prod_1^{2n} (u - u_\nu)}, \qquad \gamma(u) = \sigma_+(u)\, P_{n-1}(u), \tag{17.33}$$

where P_{n-1} is an arbitrary polynomial of degree not higher than $n - 1$.

We now wish to show conversely that the functions $\phi(u)$ and $\gamma(u)$ given in (17.33) correspond to actual solutions of (17.10) and (17.11). The function $\phi(u)$ is regular over $s \leqq \beta$ except for poles at $u = u_1, \cdots, u_{2n}$, and is $O(1/|u|)$ for large $|u|$. Similarly, $\gamma(u)$ is regular over $s \geqq -\beta$ and is $O(1/|u|)$ for large $|u|$. Thus by Plancherel's theorem, the integrals in the Mellin inversion formulas

$$\begin{aligned} f(x) &= \underset{A\to\infty}{\text{l.i.m.}}\ \frac{1}{2\pi i} \int_{-\beta - iA}^{-\beta + iA} \phi(u)\, e^{-ux}\, du, \\ g(x) &= \underset{A\to\infty}{\text{l.i.m.}}\ \frac{1}{2\pi i} \int_{-\beta - iA}^{-\beta + iA} \gamma(u)\, e^{-ux}\, du, \end{aligned} \tag{17.34}$$

have a sense, and determine functions $f(x)$ and $g(x)$ belonging to L_2. By (3.38), we may write

$$f(x)\, e^{-\lambda x} = \frac{1}{2\pi} \int_{-\infty}^{\infty} \phi(-\lambda + it)\, e^{-ixt}\, dt \qquad (\lambda \geqq \beta), \tag{17.35}$$

and by Theorem V we may conclude that $f(x)$ is equivalent to zero for negative values of x, while $g(x)$ is equivalent to zero for positive values of x. That is, (17.11) holds good.

As to (17.10), we must show that

$$\delta(x) = e^{-\beta x}\left(g(x) - f(x) + \int_{-\infty}^{\infty} K(x - y)\, f(y)\, dy\right) \tag{17.36}$$

vanishes, and in the first instance that it vanishes almost everywhere. Now, by (17.34), $e^{-\beta x}\, g(x)$ and $e^{-\beta x} f(x)$ are the Fourier transforms of

$$\frac{1}{i(2\pi)^{1/2}}\, \phi(it - \beta) \quad \text{and} \quad \frac{1}{i(2\pi)^{1/2}}\, \gamma(it - \beta).$$

Thus the integral

$$e^{-\beta x} \int_{-\infty}^{\infty} K(x - y)\, f(y)\, dy$$

in (17.36) may be transformed by Parseval's theorem into

$$\underset{A\to\infty}{\text{l.i.m.}}\ \frac{e^{-\beta x}}{2\pi i} \int_{-\beta - iA}^{-\beta + iA} \phi(u)\, \kappa(u)\, e^{-ux}\, du. \tag{17.37}$$

That is, the Fourier transform of the integral in (17.36) is

$$\frac{\phi(it - \beta)\, \kappa(it - \beta)}{i(2\pi)^{1/2}}, \tag{17.38}$$

and the Fourier transform of $\delta(x)$ is

$$(17.39)\quad \frac{1}{i(2\pi)^{1/2}}\left(-\phi(-\beta+it)+\gamma(-\beta+it)+\phi(-\beta+it)\,\kappa(-\beta+it)\right)=0,$$

and $\delta(x)$ vanishes almost everywhere. Thus the function $f(x)$ defined by (17.33) and (17.34) satisfies the integral equation (17.01) almost everywhere. It is finally easy to prove from (17.01) and the fact that $f(x)\,e^{-\beta x}$ belongs to L_2, that $f(x)\,e^{-\beta x}$ is continuous and bounded.

We now wish to discuss the asymptotic behavior of $f(x)$. We restrict ourselves to the case where $K(x)$ is even. In (17.34), if we displace the abscissa of integration from $-\beta$ to $+\beta$, we change the value of the integral by the negative sum of the residues of $\phi(u)\,e^{-ux}$ in $|\,\Re(u)\,|\leqq\alpha$. By (17.33), such a residue is of the form

$$(17.40)\quad -\,Q(x)\,e^{-u^*x},$$

where u^* denotes a zero of $1-\kappa(u)$ within the strip $|\,\Re(u)\,|\leqq\alpha$, and $Q(x)$ is a polynomial of degree less than the multiplicity of the zero. Thus

$$(17.41)\quad \begin{aligned} f(x) &= f_0(x)+r(x);\\ f_0(x) &= \Sigma\,Q(x)\,e^{-u^*x}; \qquad r(x)=\frac{1}{2\pi i}\int_{\beta-i\infty}^{\beta+i\infty}\phi(u)\,e^{-ux}\,du\,; \end{aligned}$$

where the sum f $f_0(x)$ is taken over all roots u^* in $|\,\Re(u)\,|\leqq\alpha$. The function $f_0(x)$ is a solution of the integral equation (17.04), and thus satisfies the equation

$$(17.42)\quad f_0(x)=\int_0^\infty K(x-y)\,f_0(y)\,dy+\int_{-\infty}^0 K(x-y)\,f_0(y)\,dy\,.$$

Clearly $f_0(x)=O(e^{\beta|x|})$, so that by use of our assumptions concerning K, we get

$$(17.43)\quad \int_{-\infty}^0 K(x-y)\,f_0(y)\,dy=O(e^{-\beta x})\,.$$

By (17.41), $r(x)\,e^{\beta x}$ is the Fourier transform of

$$\frac{1}{(2\pi)^{1/2}}\,\phi(\beta-it)$$

and belongs to L_2. By (17.41–3),

$$(17.44)\quad r(x)=\int_0^\infty K(x-y)\,r(y)\,dy+O(e^{-\beta x})\,.$$

The integrand is $(K(x-y)\,e^{-\beta y})\,(r(y)\,e^{\beta y})$, so that, by the Schwarz inequality, $r(x)=O(e^{-\beta x})$. Thus under the assumptions we have made in this section concerning K, we have established

THEOREM XVI. *Let 2n be the (always finite) number of zeros of*

$$(17.45)\qquad 1 - \int_{-\infty}^{\infty} K(x)\, e^{ux}\, dx \qquad [K(x) = K(|x|)],$$

counted with their proper multiplicity, in the strip $\Re(u) \leqq \alpha < 1$. *Then the maximum number of linearly independent solutions of* (17.01) *for which* $f(x) = O(e^{(\alpha+\delta)x})$, δ *being an arbitrary positive number, is exactly n. The solutions are of the form*

$$(17.46)\qquad f(x) = \Sigma\, Q(x)\, e^{-u^* x} + O(e^{-\beta x})\,,\dagger$$

where u^* *is one of the above zeros and* $Q(x)$ *is a polynomial of degree less than the multiplicity of* u^*; *and* β *is a number interior to* $(\alpha, 1)$ *such that the strips* $\alpha < \Re(u) \leqq \beta$, $-\alpha > \Re(u) \geqq -\beta$ *contain no zeros.*

Let us mention as a further property of the even kernel that among the terms in the principal part of the solution, there is always one with $\Re(u^*) \leqq 0$. That is, there is no solution of (17.01) for which $f(x) \to 0$, $x \to \infty$, for by (17.33) and $\sigma_-(u) \not\equiv 0$, $\phi(u)$ has at least one pole which does not lie in the right half-plane.

Let us now apply our general theory to some special examples. In the Lalesco equation,

$$K(x) = \lambda e^{-|x|} \qquad \text{and} \qquad \kappa(u) = \frac{2\lambda}{1 - u^2}.$$

The roots of $1 - \kappa = 0$ are $u = \pm(1 - 2\lambda)^{1/2}$ and we have $n = 1$. The representation (17.17) yields directly

$$(17.47)\qquad \sigma_+(u) = \frac{1}{u + 1}.$$

For $\lambda \leqq 0$ there are no solutions with property (17.09) and any $\alpha < 1$. For $\lambda > 0$, there is essentially only one solution. By (17.33), we have

$$(17.48)\qquad \phi(u) = \frac{u - 1}{u^2 - 1 + 2\lambda}.$$

Furthermore, by (17.41),

$$(17.49)\qquad \begin{aligned} f(x) = {} & \frac{e^{x(1-2\lambda)^{1/2}} + e^{-x(1-2\lambda)^{1/2}}}{2} + \frac{e^{x(1-2\lambda)^{1/2}} - e^{-x(1-2\lambda)^{1/2}}}{2(1 - 2\lambda)^{1/2}} \\ & + \frac{1}{2\pi i} \int_{\beta - i\infty}^{\beta + i\infty} \phi(u)\, e^{-ux}\, du\,. \end{aligned}$$

The integral here must vanish, as the abscissa of integration may be displaced arbitrarily far to the right.

† (17.09) need not be satisfied, since a multiple zero may lie on the border of the strip $|\Re(u)| \leqq \alpha$ and hence a term of the order of magnitude $x^l e^{\alpha x}$, $l < 0$, may occur.

A second, more complicated, example is that of the Milne equation. Here

$$K(x) = \frac{1}{2}\int_{|x|}^{\infty} \frac{e^{-t}}{t}\,dt, \tag{17.50}$$

so that

$$\begin{aligned}\kappa(u) &= \frac{1}{2}\int_{x=0}^{\infty}\int_{t=x}^{\infty}\frac{e^{-t}}{t}(e^{ux}+e^{-ux})\,dt\,dx \\ &= \frac{1}{2}\int_{t=0}^{\infty}\int_{x=0}^{t}\frac{e^{-t}}{t}(e^{ux}+e^{-ux})\,dx\,dt \\ &= \frac{1}{2u}\int_{0}^{\infty}\frac{e^{-(1-u)t}-e^{-(1+u)t}}{t}\,dt \\ &= \frac{1}{2u}\int_{0}^{\infty}dt\int_{1-u}^{1+u}e^{-wt}\,dw = \frac{1}{2u}\int_{1-u}^{1+u}\frac{dw}{w} = \frac{1}{2u}\log\frac{1+u}{1-u},\end{aligned} \tag{17.51}$$

where we take that branch of the logarithm which is regular in $(|\Re(u)| < 1)$ and vanishes for $u = 0$. The equation $1 - \kappa = 0$ has $u = 0$ as a double root. A little reflection will show that there are no other roots in the strip $|\Re(u)| < 1$. Thus there is only one solution with the property (17.09). If x is large, this behaves like a linear function:

$$f(x) = x + a + O(e^{-(1-\delta)x}), \tag{17.52}$$

where δ is an arbitrary positive number. We have not been able in this case to reduce the integral representation of $f(x)$ to elementary form. It is presumably a new transcendent.

In the general case, the assumption (17.09) concerning the solutions of (17.01) is quite natural, since $1 - \kappa$ may have an infinity of zeros in the strip $|\Re(u)| < 1$. In case $1 - \kappa$ has only a finite number of zeros, as for instance in the two cases just discussed, we may prove the uniqueness of our solutions under even weaker assumptions than (17.09), but we shall not pursue the matter further.

We now come to the special case where the kernel is positive.† Let $K = K(|x|) > 0$ except for a finite number of places, and let $K(0) \leqq 1$, $K(1) > 1$. Then there will be at least one positive solution of (17.01). For under these assumptions,

$$\kappa(u) = \int_0^{\infty} K(x)(e^{ux} + e^{-ux})\,dx$$

increases monotonely as u increases from 0 through real values to 1, and the equation $1 - \kappa = 0$ has just two real roots $\pm u^*(u^* \geqq 0)$ in the strip $|\Re(u)| < 1$. Thus there is a solution of (17.01) for which $f(x) = O(e^{u^*x})$ if $u^* > 0$ and $f(x) = O(x)$ if $u^* = 0$, which means that $\kappa(0) = 1$. In both cases, the solu-

† Cf. also E. Hopf, *Über lineare Integralgleichungen mit positivem Kern,* Berliner Sitzungsberichte, 1928, XVIII.

tion is positive from some point on. If it is anywhere zero or negative, it must somewhere assume its lower bound, say at $x = x_0$. Since $K > 0$,

$$f(x_0) = \int_0^\infty K(x_0 - y) f(y)\, dy \geqq f(x_0) \int_0^\infty K(x_0 - y)\, dy, \tag{17.53}$$

or

$$f(x_0)\left\{1 - \int_0^\infty K(x_0 - y)\, dy\right\} \geqq 0, \tag{17.54}$$

where we can only have equality in case $f \equiv f(x_0)$. Thus $f(x_0) = 0$ is impossible. $f(x_0) < 0$ is likewise impossible, since the integral in parentheses is less than

$$\int_{-\infty}^\infty K(t)\, dt = \kappa(0) \leqq 1.$$

Thus $f(x) > 0$ for $x \geqq 0$. For example, solution (17.52) of Milne's equation is positive.

18. **A note on the Volterra equation.** A theorem of Mercer* asserts that if $0 < \alpha < 1$, and if

$$\alpha s_n + (1 - \alpha) \frac{1}{n} \sum_1^n s_\nu \to s, \tag{18.01}$$

then

$$s_n \to s. \tag{18.02}$$

This theorem possesses generalizations of a non-trivial nature. The continuous analogue asserts that if $0 < \alpha < 1$ and if

$$\alpha s(x) + \frac{1 - \alpha}{x} \int_1^x s(y)\, dy \to s \qquad \text{as} \qquad x \to \infty, \tag{18.03}$$

then

$$s(x) \to s. \tag{18.04}$$

By a change of independent variable, this asserts that if

$$\alpha S(\xi) + (1 - \alpha) \int_0^\xi e^{\eta - \xi} S(\eta)\, d\eta \to s \qquad \text{as} \qquad \xi \to \infty, \tag{18.05}$$

* J. Mercer, *On the limits of real variants*, Proceedings of the London Mathematical Society, (2), vol. 5 (1907), pp. 206–224.

Paley and Wiener, loc. cit., Note VII, *On the Volterra equation*, Transactions of the American Mathematical Society, vol. 35, pp. 785–791.

L. L. Silverman, *On the consistency and equivalence of certain generalized definitions of the limit of a function of a continuous variable*, Annals of Mathematics, vol. 21 (1920), pp. 128–140.

then

$$S(\xi) \to s. \tag{18.06}$$

This statement is a particular case of the following theorem:

THEOREM XVII. *Let $F(x)$ be measurable and bounded over every finite range (O, A). Let $K(x)$ belong to L over $(0, \infty)$: that is,*

$$\int_0^\infty |K(\xi)| \, d\xi < \infty. \tag{18.07}$$

Let

$$F(x) + \int_0^x K(x - \xi) F(\xi) \, d\xi \to s \quad \text{as} \quad x \to \infty. \tag{18.08}$$

Then if

$$\int_0^\infty K(\xi) e^{-w\xi} \, d\xi \neq -1, \qquad \Re(w) \geqq 0, \tag{18.09}$$

we shall have

$$F(x) \to \frac{s}{1 + \int_0^\infty K(\xi) \, d\xi}. \tag{18.10}$$

Conversely, let $K(x)$ belong to L, let

$$\int_0^\infty K(\xi) \, d\xi \neq -1,$$

and let (18.08) *imply* (18.10) *for every $F(x)$ satisfying our conditions. Then* (18.09) *must be true.*

The second part of this theorem may be proved by reductio ad absurdum by putting

$$F(x) = e^{w_0 x}, \tag{18.11}$$

where

$$\int_0^\infty K(\xi) e^{-w_0 \xi} \, d\xi = -1, \qquad \Re(w_0) \geqq 0, \qquad w_0 \neq 0. \tag{18.12}$$

Then

$$\begin{aligned} \left| F(x) + \int_0^x K(x - \xi) F(\xi) \, d\xi \right| &= \left| e^{w_0 x} \int_x^\infty K(\xi) e^{-w_0 \xi} \, d\xi \right| \\ &\leqq \int_x^\infty |K(\xi)| \, d\xi \to 0. \end{aligned} \tag{18.13}$$

As (18.10) is obviously false, the second part of the theorem is proved.

The first part of Theorem XVII will appear as a corollary to a theorem concerning the Volterra integral equation of the closed cycle.

We shall use the symbol

$$A \star B(x) \tag{18.14}$$

to indicate the "Faltung" of the two functions $A(x)$, $B(x)$;

$$A \star B(x) = \int_0^x A(\xi) B(x - \xi)\, d\xi = \int_0^x A(x - \xi)\, B(\xi)\, d\xi = B \star A(x). \tag{18.15}$$

It is well known* that the (bounded and measurable) solution of the Volterra integral equation

$$G(x) = F(x) + K \star F(x) \tag{18.16}$$

is uniquely determined and given by

$$F(x) = G(x) + Q \star G(x), \tag{18.17}$$

where the resolvent kernel $Q(x)$ itself is determined from

$$Q(x) + K(x) = K \star Q(x) = Q \star K(x), \tag{18.18}$$

or else, by

$$Q(x) = \sum_{n=1}^{\infty} (-1)^n K^n(x), \qquad K^n(x) = K \star K^{n-1}(x), \qquad K'(x) = K(x). \tag{18.19}$$

We observe that the solution of (18.18) is easily obtained by using Laplace transforms.† Let us designate by

$$k(w) = \int_0^\infty K(\xi)\, e^{-w\xi}\, d\xi, \tag{18.20}$$

$$q(w) = \int_0^\infty Q(\xi)\, e^{-w\xi}\, d\xi, \tag{18.21}$$

the Laplace transforms of $K(x)$, $Q(x)$. Equation (18.19) reduces then to

$$q(w) = \frac{-k(w)}{1 + k(w)} \tag{18.22}$$

and $Q(x)$ will be found by the inversion of a Laplace integral.

The theorem in question may now be stated as follows:

THEOREM XVIII. *A necessary and sufficient condition that $Q(x)$ belong to L over $(0, \infty)$ (that is that*

$$\int_0^\infty |\, Q(\xi)\, |\, d\xi < \infty), \tag{18.23}$$

* Volterra, *Leçons sur les Equations Intégrales et les Equations Intégro-Differéntielles*, Paris, 1913.

† See, e.g., S. Bochner, *Vorlesungen über Fouriersche Integrale*, Leipzig, 1932, Chapter VII. Other references are also found there.

is that

$$k(w) = \int_0^\infty K(\xi)\, e^{-w\xi}\, d\xi \neq -1 \qquad (\Re(w) \geqq 0). \tag{18.24}$$

If this theorem holds true, the first part of Theorem XVII is immediately derived. Indeed, under the assumptions made we have

$$F(x) = G(x) + \int_0^x G(x - \xi)\, Q(\xi)\, d\xi. \tag{18.25}$$

Here $G(\xi)$ is bounded over every finite range and $\to s$ as $\xi \to \infty$. Hence $G(\xi)$ is bounded over the whole range $(0, \infty)$. Since $Q(\xi)$ is integrable over $(0, \infty)$ we may pass to the limit under the integral sign as $x \to \infty$, with the result

$$F(x) \to s + s \int_0^\infty Q(\xi)\, d\xi = s[1 + q(0)] = s[1 + k(0)]^{-1}, \tag{18.26}$$

which is precisely the desired formula (18.10).

To prove the necessity of (18.24) we observe that if (18.23) holds then $q(w)$ as well as $k(w)$ is analytic in the half-plane $\Re(w) > 0$ and continuous up to and including the boundary $\Re(w) = 0$. This implies that the denominator in the right-hand member of (18.22) does not vanish for $\Re(w) \geqq 0$ so that (18.24) holds.

The proof that (18.24) is sufficient is more difficult. We introduce the auxiliary functions

$$\phi_A(u) = \begin{cases} 1 & [|u| < A]; \\ 2 - \dfrac{|u|}{A} & [A \leqq |u| \leqq 2A]; \\ 0 & [|u| > 2A]; \end{cases} \tag{18.27}$$

and put

$$q^*(w) = \frac{-k(w)}{1 + k(w)}; \tag{18.28}$$

$$q^*(iu) = q_1(u) + q_2(u); \tag{18.29}$$

$$q_1(u) = \phi_A(u)\, q^*(iu); \qquad q_2(u) = [1 - \phi_A(u)]\, q^*(iu). \tag{18.30}$$

We wish to prove that if A is sufficiently large $q_1(u)$ and $q_2(u)$ are both Fourier transforms of functions of L.

To begin with,

$$q_1(u) = \begin{cases} \dfrac{-\phi_A(u)\, k(iu)}{\phi_{2A}(u) + \phi_{2A}(u)\, k(iu)} & \text{when } |u| < 2A; \\ 0 & \text{when } |u| \geqq 2A. \end{cases} \tag{18.31}$$

Thus $q_1(u)$ is the quotient of two functions, each the Fourier transform of a function of L, each vanishing outside a finite range, and such that the denominator function only vanishes in points interior to the region of vanishing of the numerator function. We may then appeal to a theory due to Wiener† to show that $q_1(u)$ is the Fourier transform of a function of L.

We have

$$q_2(u) = [1 - \phi_A(u)] \frac{-k(iu)[1 - \phi_{A/2}(u)]}{1 + k(iu)[1 - \phi_{A/2}(u)]}. \tag{18.32}$$

It is easy to show that this is the Fourier transform of a function of L when the same is true of

$$\begin{aligned} -k(iu)[1 - \phi_{A/2}(u)]\{1 + k(iu)[1 - \phi_{A/2}(u)]\}^{-1} \\ = \sum_{n-1}^{\infty} (-1)^n \{k(iu)[1 - \phi_{A/2}(u)]\}^n. \end{aligned} \tag{18.33}$$

Now

$$\{k(iu)[1 - \phi_{A/2}(u)]\}^n \tag{18.34}$$

is the Fourier transform of a function $h_n(x)$ for which

$$\begin{aligned} \int_{-\infty}^{\infty} | h_n(\xi) | \, d\xi \leqq \left[\int_{-\infty}^{\infty} | h_1(\xi) | \, d\xi \right]^n \\ = \left[\int_{-\infty}^{\infty} d\xi \left| K(\xi) - \frac{1}{\pi A} \int_0^{\infty} K(\eta) \frac{\cos \frac{A}{2}(\xi - \eta) - \cos A(\xi - \eta)}{(\xi - \eta)^2} \, d\eta \right| \right]^n. \end{aligned} \tag{18.35}$$

An argument of the familiar Fejér type will show that we may choose A so large that the integral in brackets is less than any given number λ in $0 < \lambda < 1$. It will follow at once that $q_2(u)$ is the Fourier transform of a function $F_2(x)$ for which

$$\int_{-\infty}^{\infty} | F_2(\xi) | \, d\xi \leqq \frac{\lambda}{1 - \lambda}. \tag{18.36}$$

Combining this with the similar result for $q_1(u)$, we see that we may write

$$q^*(iu) = - \frac{k(iu)}{1 + k(iu)} = \int_{-\infty}^{\infty} E(\xi) \, e^{-iu\xi} \, d\xi, \text{ where } \int_{-\infty}^{\infty} | E(\xi) | \, d\xi < \infty. \tag{18.37}$$

We may rewrite (18.37) as

$$\int_{-\infty}^{0} E(\xi) \, e^{-iu\xi} \, d\xi = - \int_0^{\infty} E(\xi) \, e^{-iu\xi} \, d\xi + q^*(iu). \tag{18.38}$$

Now, it is readily seen that $k(w) \to 0$ as $| w | \to \infty$, uniformly in the half-plane $\Re(w) \geqq 0$. Since by hypothesis $1 + k(w) \neq 0$ for $\Re(w) \geqq 0$, there exists a positive constant c such that

$$| 1 + k(w) | \geqq c > 0. \tag{18.39}$$

† N. Wiener, *The Fourier Integral and Certain of its Applications*, Cambridge, 1933; Lemmas 6_7, 6_{10}, 6_{18}.

Thus

(18.40)
$$-\int_0^\infty E(\xi)\, e^{-w\xi}\, d\xi + q^*(w)$$

is a function of w analytic and bounded in the right half-plane, and continuous up to and including the imaginary axis. Similarly,

(18.41)
$$\int_{-\infty}^0 E(\xi)\, e^{-w\xi}\, d\xi$$

is a function of w analytic and bounded in the left half-plane, and continuous up to and including the imaginary axis. Furthermore, the two functions are identical on the imaginary axis. By the classical argument of Riemann-Painlevé it readily results that they are parts of the same analytic function, which is thus entire and bounded. It hence reduces to a constant, and since

(18.42)
$$\int_{-\infty}^0 E(\xi)\, e^{-w\xi}\, d\xi \to 0 \text{ as } w \to \infty,$$

this constant can only be 0. Thus

(18.43)
$$q^*(w) = \frac{-\,k(w)}{1 + k(w)} = \int_0^\infty E(\xi)\, e^{-w\xi}\, d\xi.$$

On the other hand, it follows readily from (18.19) that there exists a $w_0 > 0$ such that

(18.44)
$$\frac{-\,k(w)}{1 + k(w)} = \int_0^\infty Q(\xi)\, e^{-w\xi}\, d\xi \qquad (\Re(w) > w_0);$$

and

(18.45)
$$\int_0^\infty |\, Q(\xi)\, |\, e^{-w_0\xi}\, d\xi < \infty .$$

By the uniqueness theorem for Laplace transforms we conclude that $E(x)\, e^{-wx}$ and $Q(x)e^{-wx}$ coincide almost everywhere, whence $Q(x)$ belongs to L.

In this proof, we have used the theorem of Wiener† that if a function has an absolutely convergent Fourier series and does not vanish, its reciprocal has an absolutely convergent Fourier series. P. Lévy‡ has pointed out that the same methods suffice for the following theorem: if a function $f(x)$ has an absolutely convergent Fourier series, and Φ is analytic over the range of values of $f(x)$, then $\Phi[f(x)]$ has an absolutely convergent Fourier series. By methods not essentially different from those of this paper, we may extend this theorem as follows: if $f(x)$ is the Fourier transform of a function of L, and $\Phi(u)$ is analytic over the range of values of $f(x)$, including 0, then

$$\Phi[f(x)]$$

is the Fourier transform of a function of L.

† Loc. cit., Lemma 6_{16}.

‡ P. Lévy, *Sur le convergence absolue des séries de Fourier*, Comptes Rendus de l'Académie des Sciences, vol. 196 (1933), p. 463.

19. A theorem of Hardy. Hardy* has proved the following theorem:

THEOREM XIX. *Let $f(x)$ be measurable and let*

$$f(x) = O(|x|^n e^{-x^2/2}) \quad as \quad x \to \pm\infty, \tag{19.01}$$

and let

$$g(u) = \frac{1}{(2\pi)^{1/2}} \int_{-\infty}^{\infty} f(x)\, e^{iux}\, dx. \tag{19.02}$$

If

$$g(u) = O(|u|^n e^{-u^2/2}) \text{ as } u \to \pm\infty, \tag{19.03}$$

then $f(x)$ is of the form $P(x)\, e^{-x^2/2}$ where $P(x)$ stands for a polynomial of degree not exceeding n.

We shall prove this theorem by methods analogous to those which we have already used in this chapter. We may clearly separate $f(x)$ into even and odd parts, and treat these separately. The proofs in the two cases run quite parallel, and we shall therefore confine ourselves to the even case. In this case,

$$g(u) = \left(\frac{2}{\pi}\right)^{1/2} \int_0^{\infty} f(x) \cos ux\, dx. \tag{19.04}$$

Furthermore, if $\Re(z) > -\frac{1}{2}$,

$$\begin{aligned} \left| \int_0^{\infty} g(u)\, u^{z-1/2}\, du \right| &= A \int_0^{\infty} u^{\Re(z)-1/2+n}\, e^{-u^2/2}\, du \\ &= A \int_0^{\infty} (2v)^{(\Re(z)-3/2+n)/2}\, e^{-v}\, dv \\ &= A \cdot 2^{(\Re(z)+n)/2-3/4}\, \Gamma\left(\frac{\Re(z)+n}{2} + \frac{1}{4}\right), \end{aligned} \tag{19.05}$$

where A represents various constants. Similarly,

$$\begin{aligned} \left| \int_0^{\infty} f(x)\, x^{z-1/2}\, dx \right| &\leqq \text{const. } 2^{(\Re(z)+n)/2-3/4} \\ &\quad \times \Gamma\left(\frac{\Re(z)+n}{2} + \frac{1}{4}\right), \qquad \Re(z) > -\frac{1}{2}. \end{aligned} \tag{19.06}$$

Making use of the hypothesis, we have

$$\begin{aligned} \int_0^{\infty} g(u)\, u^{z-1/2}\, du &= \lim_{\epsilon \to 0} \int_0^{\infty} e^{-\epsilon u}\, g(u)\, u^{z-1/2}\, du \\ &= \lim_{\epsilon \to 0} \int_0^{\infty} e^{-\epsilon u}\, u^{z-1/2}\, du \left(\frac{2}{\pi}\right)^{1/2} \int_0^{\infty} f(x) \cos ux\, dx \\ &= \lim_{\epsilon \to 0} \int_0^{\infty} f(x)\, dx \left(\frac{2}{\pi}\right)^{1/2} \int_0^{\infty} \cos ux\, u^{z-1/2}\, e^{-\epsilon u}\, du\,. \end{aligned} \tag{19.07}$$

* G. H. Hardy, *A theorem concerning Fourier transforms*, Journal of the London Mathematical Society, vol. 8 (1933), pp. 227–231.

However, we have

$$\begin{aligned}\int_0^\infty \cos ux\, u^{z-1/2} e^{-\epsilon u}\, du &= x^{-z-1/2} \int_0^\infty \cos y\, y^{z-1/2} e^{-\epsilon y/x}\, dy \\ &= x^{-z-1/2} \int_0^\infty \tfrac{1}{2} [e^{-y(-i+\epsilon/x)} + e^{-y(i+\epsilon/x)}]\, y^{z-1/2}\, dy .\end{aligned} \tag{19.08}$$

By using the fact that the integral around a closed contour is zero, we have

$$\begin{aligned}\int_0^\infty e^{-y(-i+\epsilon/x)} y^{z-1/2}\, dy &= \int_0^{\infty(i+\epsilon/x)} e^{-y(-i+\epsilon/x)} y^{z-1/2}\, dy \\ &= \left(-i+\frac{\epsilon}{x}\right)^{-z-1/2} \int_0^\infty e^{-w} w^{z-1/2}\, dw \\ &= \left(-i+\frac{\epsilon}{x}\right)^{-z-1/2} \Gamma(z+\tfrac{1}{2}) .\end{aligned} \tag{19.081}$$

Similarly,

$$\int_0^\infty e^{-y(i+\epsilon/x)} y^{z-1/2}\, dy = \Gamma\left(z+\frac{1}{2}\right)\left(i+\frac{\epsilon}{x}\right)^{-z-1/2} . \tag{19.082}$$

Thus

$$\begin{aligned}&\int_0^\infty \cos ux\, u^{z-1/2} e^{-\epsilon u}\, du \\ &= x^{-z-1/2}\Gamma\left(z+\frac{1}{2}\right)\frac{1}{2}\left[\left(-i+\frac{\epsilon}{x}\right)^{-z-1/2}+\left(i+\frac{\epsilon}{x}\right)^{-z-1/2}\right] \\ &= x^{-z-1/2}\Gamma\left(z+\frac{1}{2}\right)\exp\left\{-\frac{(z+\frac{1}{2})}{2}\log\left(1+\frac{\epsilon^2}{x^2}\right)\right\}\cos\left\{\left(z+\frac{1}{2}\right)\tan^{-1}\frac{x}{\epsilon}\right\}.\end{aligned} \tag{19.083}$$

For purely imaginary z and $x > 0$ we have therefore

$$\left|\int_0^\infty \cos ux\, u^{z-1/2} e^{-\epsilon u}\, du\right| \leqq x^{-1/2} e^{\pi|z|/2}\, |\Gamma(z+\tfrac{1}{2})| . \tag{19.09}$$

Thus by dominated convergence we have

$$\begin{aligned}&\lim_{\epsilon\to 0}\int_0^\infty f(x)\, dx \left(\frac{2}{\pi}\right)^{1/2}\int_0^\infty \cos ux\, u^{z-1/2} e^{-\epsilon u}\, du \\ &= \int_0^\infty f(x)\, x^{-z-1/2}\, dx \left(\frac{2}{\pi}\right)^{1/2}\Gamma\left(z+\frac{1}{2}\right)\cos\frac{\pi}{2}\left(z+\frac{1}{2}\right).\end{aligned} \tag{19.10}$$

Using (19.07) we have

$$\int_0^\infty g(u)\, u^{z-1/2}\, du = \int_0^\infty f(x)\, x^{-z-1/2}\, dx \left(\frac{2}{\pi}\right)^{1/2}\Gamma\left(z+\frac{1}{2}\right)\cos\frac{\pi}{2}\left(z+\frac{1}{2}\right). \tag{19.11}$$

In particular,

$$\int_0^\infty e^{-u^2/2} u^{z-1/2}\, du = \int_0^\infty e^{-x^2/2} x^{-z-1/2}\, dx \left(\frac{2}{\pi}\right)^{1/2}\cos\frac{\pi}{2}\left(z+\frac{1}{2}\right)\Gamma\left(z+\frac{1}{2}\right). \tag{19.12}$$

However,

$$\int_0^\infty e^{-u^2/2} u^{z-1/2}\, du = 2^{z/2-3/4}\,\Gamma\left(\frac{z}{2}+\frac{1}{4}\right). \tag{19.13}$$

Thus

$$\frac{2^{-z/2}\int_0^\infty g(u)\, u^{z-1/2}\, du}{\Gamma\left(\frac{z}{2}+\frac{1}{4}\right)} = \frac{2^{z/2}\int_0^\infty f(x)\, x^{-z-1/2}\, dx}{\Gamma\left(-\frac{z}{2}+\frac{1}{4}\right)}. \tag{19.14}$$

Let us call this function $F(z)$. It is clear by (19.05) and (19.06) that it is entire, and in the right half-plane,

$$|\, F(z)\, | \leqq \text{const.}\, |\, z\, |^{n/2} \exp\, (\pi\, |\, \Im(z)\, |\, /4)\, . \tag{19.15}$$

In the same way we may prove that (19.15) holds in the left half-plane, and hence in the whole plane.

Let us now consider $F(z)$ on the positive imaginary axis. Here

$$\begin{aligned} |\, F(iy)\, | &= \frac{\left|\int_0^\infty g(u)\, u^{iy-1/2}\, du\right|}{\left|\Gamma\left(\frac{iy}{2}+\frac{1}{4}\right)\right|} \sim A\, y^{-1/4}\, e^{\pi y/4} \left|\int_0^\infty g(u)\, u^{iy-1/2}\, du\right| \\ &\sim A\, y^{-1/4} \left| \int_0^{\infty \exp(-\pi i/4 + \epsilon i)} g(u) \exp\left(i\left(\frac{\pi}{4}-\epsilon\right)\right) u^{iy-1/2}\, du \right| e^{\epsilon y}. \end{aligned} \tag{19.16}$$

Now,

$$\begin{aligned} |\, g(ue^{i\theta})\, | &= \left| \frac{1}{2\pi} \int_{-\infty}^\infty f(x) \exp\, (ixue^{i\theta})\, dx \right| \\ &\leqq A \left| \int_{-\infty}^\infty |\, x\,|^n \exp\left(-\frac{x^2}{2} + ixue^{i\theta}\right) dx \right| \\ &\leqq A \left| \int_{-\infty}^\infty |\, x\,|^n \exp\left(-\frac{1}{2}\, (x - iue^{i\theta})^2\right) dx \exp\left(\frac{-u^2 e^{2i\theta}}{2}\right) \right| \\ &\leqq \left[A \int_0^\infty |\, x + iue^{i\theta}\, |^n e^{-x^2/2}\, dx + A \int_0^\infty |\, x - iue^{i\theta}\, |^n e^{-x^2/2}\, dx \right] \\ &\qquad \times \exp\left[-\frac{u^2}{2} \cos\left(\frac{\pi}{2} - 2\epsilon\right) \right] \\ &\leqq A(1 + |\, u\,|^n) \exp\left(-O(1)\, \frac{u^2}{2}\right) \quad \left[-\frac{\pi}{4} + \epsilon \leqq \theta \leqq \frac{\pi}{4} - \epsilon \right]. \end{aligned} \tag{19.17}$$

Thus by (19.16) and (19.17),

$$|\, F(iy)\, | \sim \text{const.}\ y^{-1/4} \left| \int_0^\infty g\left(u \exp\left(i\left(\frac{\pi}{4}-\epsilon\right)\right)\right) u^{iy-1/2}\, du \right| e^{\epsilon y}, \tag{19.18}$$

where the integration is taken along the real axis. Hence

$$|F(iy)| = O(e^{\epsilon y}); \qquad |F(-iy)| = O(e^{\epsilon y}) \tag{19.19}$$

for all ϵ. Thus by the Phragmén-Lindelöf theorem and (19.15),

$$F(z) = O(e^{\epsilon_1 z}) \tag{19.20}$$

if $\epsilon_1 > \epsilon$. It follows that either $F(z)$ must be a polynomial, or it must have an infinite set of zeros, for otherwise it would be of the form e^{az}, which contradicts (19.18). Thus if $F(z)$ is not a polynomial we may remove more than $n/2 + 1$ of these zeros and obtain an entire function $G(z)$ which belongs to L_2 along the real axis, and satisfies

$$G(z) = O(e^{\epsilon |z|}) \tag{19.21}$$

for all ϵ. This is, however, impossible, by Theorem XI. Thus $F(z)$ is a polynomial, and as such it cannot, by (19.15), be of degree higher than $n/2$. Hence

$$\int_0^\infty f(x)\, x^{-z-1/2}\, dx = F(z)\, \Gamma\left(\frac{z}{2} - \frac{1}{4}\right) 2^{z/2}, \tag{19.22}$$

and over every finite range,

$$\begin{aligned} f(x) &= \operatorname*{l.i.m.}_{A\to\infty} \frac{1}{(2\pi)^{1/2}} \int_{-A}^{A} F(iu)\, \Gamma\left(\frac{iu}{2} - \frac{1}{4}\right) 2^{iu/2}\, x^{iu}\, du \\ &= \sum_0^{[n/2]} d_k \left(x \frac{d}{dx}\right)^k e^{-x^2/2} \\ &= \sum_0^{[n/2]} c_k\, x^{2k}\, e^{-x^2/2}. \end{aligned} \tag{19.23}$$

Here we make use of a mathematical induction from the fact that

$$x \frac{d}{dx}(x^n e^{-x^2/2}) = (nx^n - x^{n+2})\, e^{-x^2/2}. \tag{19.24}$$

CHAPTER V

ENTIRE FUNCTIONS OF THE EXPONENTIAL TYPE

20. Classical theorems concerning entire functions. Let $f(z)$ be an entire function: that is to say, a function without singularities in the entire plane. It is said to be of finite order if there is a positive number A such that, as $z = re^{i\theta}$ tends to infinity,

$$f(z) = O(e^{r^A}). \tag{20.01}$$

The lower bound ρ of numbers A for which this is true is called the *order* of the function. In this book we shall devote most of our attention to functions $f(z)$ of order 1; that is, to functions for which if $\epsilon > 0$,

$$f(z) = O(e^{r^{1+\epsilon}}), \tag{20.02}$$

and in particular, to the narrower class of functions of exponential type, for which there exists an A such that

$$f(z) = O(e^{rA}). \tag{20.03}$$

In these cases we may write

$$f(z) = z^m e^{cz} B \prod_1^\infty \left(1 - \frac{z}{\lambda_n}\right) e^{z/\lambda_n}, \tag{20.04}$$

by a well known theorem of Hadamard.* In particular, if $f(z)$ is even, we may multiply the factors of (20.04) by pairs, and obtain

$$f(z) = z^{2m} B \prod_1^\infty \left(1 - \frac{z^2}{\lambda_n^2}\right); \tag{20.05}$$

and if $f(z)$ is odd,

$$f(z) = z^{2m+1} B \prod_1^\infty \left(1 - \frac{z^2}{\lambda_n^2}\right). \tag{20.06}$$

If $f(z)$ is even, we have

$$\frac{\log\{|f(z)|\,|z|^{-2m}\}}{|z|} = \frac{\log|B|}{|z|} + \sum_1^\infty \frac{1}{|z|} \log\left(1 + \frac{|z|^2}{|\lambda_n|^2}\right). \tag{20.07}$$

Let us put $\lambda(r)$ for the number of λ_n's not exceeding r in modulus. Then

$$\begin{aligned} \frac{\log\{|f(z)|\,|z^{-2m}|\} - \log|B|}{|z|} &\leqq \frac{1}{|z|}\int_0^\infty \log\left(1 + \frac{|z|^2}{u^2}\right) d\lambda(u) \\ &= \lim_{A\to\infty} \left\{ \frac{1}{|z|} \log\left(1 + \frac{|z|^2}{A^2}\right)\lambda(A) + \frac{2}{|z|}\int_0^A \frac{\lambda(u)\dfrac{|z|^2}{u^2}}{1 + \dfrac{|z|^2}{u^2}}\, du \right\}. \end{aligned} \tag{20.08}$$

* E. C. Titchmarsh, loc. cit., p. 250.

It follows that

$$\overline{\lim_{z\to\infty}} \frac{\log|f(z)|}{|z|} \leqq \pi \overline{\lim_{u\to\infty}} \frac{\lambda(u)}{u}. \tag{20.09}$$

On the other hand, by Jensen's theorem, if

$$\lim_{z\to 0} f(z)\, z^{-m} = B > 0, \tag{20.10}$$

we have

$$\begin{aligned}\int_0^r \frac{2\,\lambda(x)}{x}\,dx &= 2\sum_1^n \log\frac{r}{|\lambda_k|} = 2\log\frac{r^n}{|\lambda_1\lambda_2\cdots\lambda_n|}\\ &= \frac{1}{2\pi}\int_0^{2\pi}\log|f(re^{i\theta})\,r^{-2m}\,e^{-2im\theta}|\,d\theta\log|B|,\end{aligned} \tag{20.11}$$

where $n = \lambda(r)$. Thus

$$\overline{\lim_{r\to\infty}} \frac{1}{r}\int_0^r \frac{2\,\lambda(x)}{x}\,dx = \overline{\lim_{r\to\infty}} \frac{1}{2\pi r}\int_0^{2\pi}\log|f(re^{i\theta})|\,d\theta \leqq \overline{\lim_{z\to\infty}} \frac{\log|f(z)|}{z}, \tag{20.12}$$

and

$$\overline{\lim_{r\to\infty}} \frac{1}{r}\int_{r/2}^r \frac{2\,\lambda(x)}{x}\,dx \leqq \overline{\lim_{z\to\infty}} \frac{\log|f(z)|}{|z|}. \tag{20.13}$$

Since $\lambda(u)$ is increasing, this gives us

$$\frac{1}{\log z}\,\overline{\lim_{r\to\infty}} \frac{1}{r}\int_r^{2r}\lambda(r)\frac{dx}{x} = \overline{\lim_{u\to\infty}} \frac{\lambda(u)}{u} \leqq \frac{1}{\log 2}\,\overline{\lim_{z\to\infty}} \frac{\log|f(z)|}{|z|}. \tag{20.14}$$

Combining this with (20.09) we see that

THEOREM XX. *If $f(x)$ is an even entire function of order* 1, *$\lambda(u)/u$ is bounded at infinity when and only when $\log|f(z)|/z$ is bounded at infinity, $\lambda(u)$ being defined as in the lines just following* (20.07).

A chief purpose of this chapter is to prove the following much deeper theorem of the same character:

THEOREM XXI. *Let $f(z)$ be an even entire function and equal to* 1 *at the origin. Let*

$$\int_0^\infty \frac{\log^+|f(x)|\,dx}{x^2} < \infty, \tag{20.15}$$

the integral being taken along the axis of reals, and let

$$\overline{\lim_{z\to\infty}} \frac{1}{|z|}\log|f(z)| < \infty. \tag{20.16}$$

Let the zeros of $f(x)$ be $\pm z_n$ and let

$$f_1(z) = \prod_1^\infty\left(1 - \frac{z^2}{|z_n|^2}\right).$$

Then

(20.17)
$$\lim_{u\to\infty} \frac{\lambda(u)}{u} = A$$

exists, and

(20.18)
$$A = -\frac{1}{\pi^2}\int_0^\infty \frac{\log f_1(x)\,dx}{x^2}.$$

This will in particular be the case if

(20.19)
$$\int_0^\infty |f(x)|^2\,dx < \infty,$$

the integral being taken along the real axis. We shall have

(20.20)*
$$f(z) = \int_0^{\pi L} \cos u z\,\phi(u)\,du \qquad (L > 0),$$

where $\phi(u)$ is not equivalent to zero over the whole of any interval $(\pi L - \epsilon, \pi L)$.

(20.04) may be replaced by

(20.21)
$$\int_0^\infty \log^+ m_f(r)\,\frac{dr}{r^2} < \infty,$$

where $m_f(r)$ is

$$\min_{0\leqq\theta\leqq 2\pi} |f(re^{i\theta})|.$$

21. A Tauberian theorem concerning entire functions. In the proof of Theorem XXI, the following theorem occupies an important place:

THEOREM XXII. *Let $\{\lambda_n\}$ be a monotone sequence of positive numbers such that the series $\sum_1^\infty \lambda_n^{-2}$ converges. We set*

(21.01)
$$\phi(z) = \prod_{\nu=1}^{\infty}\left(1 - \frac{z^2}{\lambda_\nu^2}\right).$$

Then the statements

(21.02)
$$\log \phi(iy) \sim \pi A\,|y| \quad \textit{as} \quad y \to \pm\infty$$

and

(21.03)
$$\int_{-\infty}^{\infty} \log|\phi(x)|\,x^{-2}\,dx = -\pi^2 A$$

are completely equivalent.

* Cf. (20.11) and Theorem X.

To prove this, let us note that if $\lambda(t)$ is the number of λ_n's not exceeding t, and $y > 0$,

$$(21.04)\qquad \begin{aligned}(\pi y)^{-1} \log \phi(iy) &= (\pi y)^{-1} \sum_{\nu=1}^{\infty} \log \left(1 + \frac{y^2}{\lambda_n^2}\right) \\ &= (\pi y)^{-1} \int_0^{\infty} \log \left(1 + \frac{y^2}{t^2}\right) d\lambda(t).\end{aligned}$$

Similarly,

$$(21.05)\qquad \begin{aligned}-\pi^{-2} \int_{-y}^{y} \log |\phi(x)| x^{-2}\, dx &= -2\pi^{-2} \int_0^y x^{-2}\, dx \int_0^{\infty} \log \left| 1 - \frac{x^2}{t^2} \right| d\lambda(t) \\ &= -2\pi^{-2} \int_0^{\infty} d\lambda(t) \int_0^y \log \left| 1 - \frac{x^2}{t^2} \right| x^{-2}\, dx \\ &= -2\pi^{-2} y^{-1} \int_0^{\infty} d\lambda(t)\, \frac{y}{t} \\ &\qquad \times \int_0^{y/t} \log |1 - s^2|\, s^{-2}\, ds.\end{aligned}$$

Expressions (21.04) and (21.05) are both of the form

$$(21.06)\qquad \frac{1}{y} \int_0^{\infty} N\left(\frac{t}{y}\right) d\lambda(t),$$

where $\lambda(t)$ is a monotone increasing function. In (21.04) we have

$$(21.07)\qquad N(\lambda) = N_1(\lambda) = \frac{1}{\pi} \log \left(1 + \frac{1}{\lambda^2}\right),$$

while in (21.05)

$$(21.08)\qquad \begin{aligned}N(\lambda) = N_2(\lambda) &= -\frac{2}{\pi^2 \lambda} \int_0^{1/\lambda} \log |1 - x^2|\, x^{-2}\, dx \\ &= \frac{2}{\pi^2} \left\{ \log \left| 1 - \frac{1}{\lambda^2} \right| + \frac{1}{\lambda} \log \left| \frac{1 + \lambda}{1 - \lambda} \right| \right\}.\end{aligned}$$

The function $N_1(\lambda)$ is positive and monotone decreasing. This is also true of $N_2(\lambda)$, since

$$(21.09)\qquad N_2'(\lambda) = -\frac{2}{\pi^2 \lambda^2} \log \left| \frac{1 + \lambda}{1 - \lambda} \right| < 0.$$

The following properties are easily established, where $N(\lambda)$ stands for either of $N_1(\lambda)$, $N_2(\lambda)$:

$$(21.10)\qquad N(\lambda) = \begin{cases} O\,(\log 1/\lambda) \text{ as } \lambda \to 0; \\ O\,(1/\lambda^2) \text{ as } \lambda \to \infty\,; \end{cases}$$

$$(21.11)\qquad \sum_{k=-\infty}^{\infty} \max_{2^k \leqq \lambda \leqq 2^{k+1}} \lambda\, N(\lambda) < \infty, \qquad N(\lambda) > 0.$$

Furthermore,

$$\int_0^\infty N_1(\lambda)\,\lambda^{it}\,d\lambda = \frac{1}{\pi(it+1)}\int_0^\infty \frac{\mu^{(it-1)/2}}{1+\mu}\,d\mu = \frac{1}{(it+1)\cosh\dfrac{\pi t}{2}} \tag{21.12}$$

and

$$\begin{aligned}\int_0^\infty N_2(\lambda)\,\lambda^{it}\,d\lambda &= \frac{2}{\pi^2(it+1)}\int_0^\infty \lambda^{it-1}\log\left|\frac{1+\lambda}{1-\lambda}\right|d\lambda \\ &= \frac{2}{\pi^2(it+1)}\sum_{k=-\infty}^{\infty}\frac{1}{\left(\dfrac{t}{2}\right)^2+\left(k+\dfrac{1}{2}\right)^2} = \frac{2\tan\dfrac{\pi it}{2}}{\pi it(it+1)}.\end{aligned} \tag{21.13}$$

Thus it follows that if t is real,

$$\int_0^\infty N(\lambda)\,\lambda^{it}\,d\lambda \neq 0, \tag{21.14}$$

and that

$$\int_0^\infty N(\lambda)\,d\lambda = 1. \tag{21.15}$$

Finally, as $y \to 0$,

$$\lim \frac{1}{y}\int_0^\infty N_1\left(\frac{t}{y}\right)d\lambda(t) = \lim\,(\pi y)^{-1}\log\phi(iy) = 0, \tag{21.16}$$

and

$$\lim \frac{1}{y}\int_0^\infty N_2\left(\frac{t}{y}\right)d\lambda(t) = \lim\left(-\pi^2\int_{-y}^{y}\log|\phi(x)|\,x^{-2}\,dx\right) = 0. \tag{21.17}$$

Thus either of the statements

$$\frac{1}{y}\int_0^\infty N_i\left(\frac{t}{y}\right)d\lambda(t) \to A \qquad \text{as} \qquad y\to\infty \qquad (i=1,2) \tag{21.18}$$

implies the boundedness of the corresponding integral

$$\frac{1}{y}\int_0^\infty N_i\left(\frac{t}{y}\right)d\lambda(t) \tag{21.19}$$

over the range $(0, \infty)$. A direct application of a Tauberian theorem of Wiener* then shows that statements (21.18) are completely equivalent, which is precisely the result of Theorem XXII.

* N. Wiener, *Tauberian theorems*, Annals of Mathematics, (2), vol. 33 (1932), pp. 1–100; Theorem XI, p. 30.

We now proceed to the proof of Theorem XXI. By Theorem XX, if we replace each pair of zeros of $f(x)$, which are the negatives of each other, by another pair of the same absolute value but real, changing in effect

$$f(z) = \prod_1^\infty \left(1 - \frac{z^2}{z_\nu^2}\right) \tag{21.20}$$

into

$$f_1(z) = \prod_1^\infty \left(1 - \frac{z^2}{|z_\nu^2|}\right), \tag{21.21}$$

we certainly do not affect the truth of (20.14), as for real x,

$$\log \left| 1 - \frac{x^2}{|z_\nu^2|} \right| \leqq \log \left| 1 - \frac{x^2}{z_\nu^2} \right|, \tag{21.22}$$

nor do we affect the truth of (20.19). If we put $\lambda(u)$ for the number of z_ν's not exceeding u in modulus, it follows that $\lambda(u)$ is not affected by this transformation, so that by Theorem XX, (20.131) remains true. Thus it is legitimate to replace $f(z)$ by $f_1(z)$ in the demonstration of the theorem, and to assume that the z_n are all real and positive. We shall suppose that

$$0 < z_1 \leqq z_2 \leqq \cdots, \tag{21.23}$$

where obviously

$$\sum_1^\infty |z_\nu|^{-2} < \infty. \tag{21.24}$$

We have

$$\frac{\lambda(u)}{u} = \frac{1}{u}\int_0^u d\lambda(v) = \frac{1}{u}\int_0^\infty K_1\left(\frac{v}{u}\right) d\lambda(v), \tag{21.25}$$

and

$$\begin{aligned} \frac{1}{\pi y} \log f_1(iy) &= \frac{1}{\pi y} \sum_1^\infty \log\left(1 + \frac{y^2}{|z_\nu^2|}\right) \\ &= \frac{1}{\pi y}\int_0^\infty \log\left(1 + \frac{y^2}{v^2}\right) d\lambda(v) = \frac{1}{y}\int_0^\infty K_2\left(\frac{v}{y}\right) d\lambda(v), \end{aligned} \tag{21.26}$$

where (cf. (21.12))

$$\begin{aligned} &K_1(u) \geqq 0; \qquad K_2(u) \geqq 0; \qquad \int_0^\infty K_1(v)\, v^{iw}\, dv = \frac{1}{iw+1} \neq 0; \\ &\int_0^\infty K_1(v)\, dv = 1; \quad \int_0^\infty K_2(v)\, v^{iw}\, dv = \frac{1}{(iw+1)\cosh(\pi w/2)} \neq 0; \\ &\int_0^\infty K_2(v)\, dv = 1. \end{aligned} \tag{21.27}$$

Thus by another theorem of Wiener,*

$$\frac{1}{y}\log f_1(iy) \sim A\pi \tag{21.28}$$

and

$$\lambda(u) \sim Au \tag{21.29}$$

are completely equivalent. This is also a result of Titchmarsh.† Thus all that we have to establish is (21.28), or by Theorem XXII,

$$\int_{-\infty}^{\infty} \log |f_1(x)|\, x^{-2}\, dx = -\pi^2 A\,. \tag{21.30}$$

We already know that

$$\log^+ |f_1(w)| = O(|w|) \tag{21.31}$$

and

$$\int_{-\infty}^{\infty} \log^+ |f_1(u)|\, u^{-2}\, du < \infty\,, \tag{21.32}$$

as follows by a direct computation. It follows from (21.31) that the ratio $\lambda(t)/t$ is bounded. Hence by (21.05) and (21.09),

$$\begin{aligned}\int_{-y}^{y} \log |f_1(u)|\, u^{-2}\, du &= -\frac{\pi^2}{y}\int_0^{\infty} N_2\left(\frac{t}{y}\right) d\lambda(t) \\ &= \frac{\pi^2}{y}\int_0^{\infty} \lambda(t)\, d_t\, N_2\left(\frac{t}{y}\right) = -2\int_0^{\infty} t^{-2}\lambda(t)\log\left|\frac{y+t}{y-t}\right| dt \\ &= O\left\{\int_0^{y}\frac{dt}{t}\log\left|\frac{1+t/y}{1-t/y}\right| + \int_y^{\infty}\frac{dt}{t}\log\left|\frac{1+y/t}{1-y/t}\right|\right\} = O(1).\end{aligned} \tag{21.33}$$

Combining this with (21.32), we obtain

$$\int_{-y}^{y} \log^- |f_1(u)|\, u^{-2}\, du = O(1)\,, \tag{21.34}$$

the integral being taken along the real axis. Thus we have

$$\int_{-\infty}^{\infty} \log^- |f_1(u)|\, u^{-2}\, du < \infty\,. \tag{21.35}$$

This and (21.32) together yield (21.30). We shall clearly have

$$\lim_{u\to\infty}\frac{\lambda(u)}{u} = A\,, \tag{21.36}$$

from which (20.16) follows at once.

* N. Wiener, *A new method in Tauberian theorems*, Journal of Mathematics and Physics, Massachusetts Institute of Technology, vol. 7 (1928), pp. 161–184. The theorem is not there stated in the form applicable to Stieltjes integrals, but the proof is substantially independent of that form.

† E. C. Titchmarsh, *On integral functions with real negative zeros*, Proceedings of the London Mathematical Society, (2), vol. 26 (1927), pp. 185–200.

Let us return to $f(x)$, in the general case discussed in the hypothesis of Theorem XXI, where not all the roots are real. By (20.13)

$$\lambda(u) = O(u). \tag{21.37}$$

On the other hand,

$$\begin{aligned}\int_0^\infty \frac{\log^+ |f(x)|}{x^2}\,dx &= \int_0^\infty \frac{\log^+ \left|\prod_1^\infty \left(1 - \frac{x^2}{z_\nu^2}\right)\right| dx}{x^2} \\ &\geqq \int_0^\infty \frac{\log^+ \left|\prod_1^\infty \left(1 - \frac{x^2}{|z_\nu^2|}\right)\right| dx}{x^2} = \int_0^\infty \frac{\log^+ |f_1(x)|\,dx}{x^2}.\end{aligned} \tag{21.38}$$

It follows that

$$A = \lim_{u\to\infty} \frac{\lambda(u)}{u} \leqq \frac{1}{\pi^2}\int_0^\infty \frac{|\log|f_1(x)||\,dx}{x^2} \tag{21.39}$$

exists.

22. A condition that the roots of an entire function be real. We wish to prove

THEOREM XXIII. *Let $f(z)$ be an even entire function of order not exceeding 1, and let the number of its roots $\pm z_n$ within a circle of radius r about the origin be $2\lambda(r)$. Let*

$$\lambda(r) \sim Br \qquad (r \to \infty). \tag{22.01}$$

Then all the roots of $f(z)$ will be real, when and only when

$$B = -\frac{1}{\pi^2}\int_{-\infty}^\infty \frac{\log|f(x)|\,dx}{x^2}. \tag{22.02}$$

By Theorem XXII and (21.28–9), all that we have to prove is that

$$\int_{-\infty}^\infty \left[\log\left|\prod_1^\infty \left(1 - \frac{u^2}{z_\nu^2}\right)\right| - \log\left|\prod_1^\infty \left(1 - \frac{u^2}{|z_\nu^2|}\right)\right|\right] u^{-2}\,du = 0 \tag{22.03}$$

when and only when every z_ν is real. However, if any z_ν is not real, we shall have

$$\log\left|\prod_1^\infty \left(1 - \frac{u^2}{z_\nu^2}\right)\right| - \log\left|\prod_1^\infty \left(1 - \frac{u^2}{|z_\nu^2|}\right)\right| = \sum_{\nu=1}^\infty \log\frac{|z_\nu^2 - u^2|}{||z_\nu^2| - u^2|} > 0 \tag{22.04}$$

for real u, which is incompatible with (22.03).

23. A theorem on the Riemann zeta function. We proceed to prove

THEOREM XXIV. *If the λ_n's are real and positive, if the series $\Sigma\, 1/\lambda_n^2$ converges, and if*

$$\phi(w) = \prod_{\nu=1}^\infty \left(1 - \frac{w^2}{\lambda_\nu^2}\right), \tag{23.001}$$

then the statements

$$\log \phi(iy) \sim \pi A \,|y|\, \log |y| \quad \textit{as } y \to \infty \tag{23.01}$$

and

$$\int_{-y}^{y} \log |\phi(x)|\, x^{-2}\, dx \sim -\pi^2 A \log |y| \tag{23.02}$$

are completely equivalent.

Let $y > 0$, and let us use the kernels $N_1(\lambda)$, $N_2(\lambda)$ of a previous paragraph. Then (23.01) and (23.02) may be replaced respectively by

$$(y \log y)^{-1} \int_0^\infty N\left(\frac{t}{y}\right) d\lambda(t) \to A\,; \qquad N = N_1(\lambda),\quad N_2(\lambda)\,. \tag{23.03}$$

We now observe that either of the statements (23.03) implies

$$\lambda(y) = O(y \log y)\,. \tag{23.04}$$

Indeed if (23.03) is satisfied then

$$\begin{aligned} O(1) &\geqq (y \log y)^{-1} \int_0^y N\left(\frac{t}{y}\right) d\lambda(t) \\ &= N(1)\,(y \log y)^{-1}\,\lambda(y) - (y \log y)^{-1} \int_0^y \lambda(t)\, d_t N\left(\frac{t}{y}\right) \\ &> N(1)\,\lambda(y)\,(y \log y)^{-1}, \end{aligned} \tag{23.05}$$

since $N(\lambda)$ is positive and decreasing. Next we prove that under condition (23.04), (23.03) is equivalent to

$$\frac{1}{y}\int_0^\infty N\left(\frac{t}{y}\right) d\Lambda^*(t) \to A\,; \qquad N(\lambda) = N_1(\lambda)\,,\quad N_2(\lambda)\,; \tag{23.06}$$

where as a Cauchy principal value,

$$\Lambda^*(y) = \int_0^y (\log t)^{-1}\, d\lambda(t)\,. \tag{23.07}$$

We can assume without essential restriction that $\lambda(t)$ is continuous at 1. It is readily seen from (23.04) and (23.07) that $\Lambda^*(y)$ vanishes for sufficiently small y, while

$$\Lambda^*(y) = O(y) \text{ as } y \to \infty\,. \tag{23.08}$$

Now the difference between the left-hand members of (23.06) and (23.03) is equal to

$$\begin{aligned} I(y) &= \frac{1}{y}\int_0^\infty N\left(\frac{t}{y}\right)\left(\frac{1}{\log t} - \frac{1}{\log y}\right) d\lambda(t) \\ &= (y \log y)^{-1} \int_0^\infty N\left(\frac{t}{y}\right) \log \frac{y}{t}\, d\Lambda^*(t) \\ &= -(y \log y)^{-1} \int_0^\infty \Lambda^*(t)\, d_t\left[N\left(\frac{t}{y}\right) \log \frac{y}{t}\right] \\ &= O\left\{(\log y)^{-1} \int_0^\infty t\, \frac{d}{dt}\left[N(t) \log \frac{1}{t}\right] dt\right\} = O\left(\frac{1}{\log y}\right), \end{aligned} \tag{23.09}$$

which tends to zero as $y \to \infty$ or $y \to 0$. The same theorem of Wiener which was applied in the proof of Theorem XXII shows immediately the equivalence of the two statements (23.06): hence the two statements (23.03) are equivalent, as are (23.01) and (23.02).

In order to apply Theorem XXIV to the theory of the Riemann zeta function, we introduce

$$\begin{aligned}\Xi(z) &= \xi\left(\frac{1}{2} + iz\right) \\ &= \frac{1}{2}\left(\frac{1}{2} + iz\right)\left(\frac{1}{2} - iz\right) \pi^{-1/4-iz/2}\,\Gamma\left(\frac{1}{4} + \frac{iz}{2}\right)\zeta\left(\frac{1}{2} + iz\right).\end{aligned} \tag{23.10}$$

It is known that $\Xi(z)$ is an entire function, is even, and has all its zeros in the strip $|\Im(z)| < \frac{1}{2}$.* Moreover

$$\log \Xi(iy) = O(y) + \log \Gamma(y/2) \sim \tfrac{1}{2}\, y \log y; \tag{23.11}$$

$$\Xi(z) = c \prod_{\nu=1}^{\infty} \left(1 - \frac{z^2}{z_\nu^2}\right), \qquad \sum_{1}^{\infty} |z_\nu|^{-2} < \infty, \qquad c = \Xi(0). \tag{23.12}$$

We set

$$z_\nu = z'_\nu + iz''_\nu; \qquad z'_\nu > 0; \qquad |z''_\nu| < \tfrac{1}{2}; \qquad |z_\nu| = \lambda_\nu. \tag{23.13}$$

Let us put

$$H(z) = c \prod_{\nu=1}^{\infty} \left(1 - \frac{z^2}{\lambda_\nu^2}\right). \tag{23.14}$$

Along the imaginary axis we have

$$\begin{aligned}\log \frac{|H(iy)|}{|\Xi(iy)|} &= -\sum_{\nu=1}^{\infty} \log \left|\frac{z_\nu^2 + y^2}{\lambda_\nu^2 + y^2}\right| = -\sum_{\nu=1}^{\infty} \log \left|1 + \frac{z_\nu^2 - \lambda_\nu^2}{\lambda_\nu^2 + y^2}\right| \\ &= O(1).\end{aligned} \tag{23.15}$$

Thus if $y > 0$, by (23.11)

$$\log H(iy) \sim \tfrac{1}{2}\, y \log y, \tag{23.16}$$

and by Theorem XXIV,

$$(\log y)^{-1} \int_{-y}^{y} \log |c^{-1} H(x)|\, x^{-2}\, dx \to -\frac{\pi}{2} \text{ as } y \to \infty. \tag{23.17}$$

Again, on the real axis,

$$\left|1 - \frac{x^2}{z_\nu^2}\right| \geqq \left|1 - \frac{x^2}{\lambda_\nu^2}\right|, \tag{23.18}$$

* Cf. A. E. Ingham, *The Distribution of Primes*, Cambridge Tract in Mathematics and Mathematical Physics, No. 30, chapter III, §7.

whence

$$\log \left| \frac{1 - x^2/z_\nu^2}{1 - x^2/\lambda_\nu^2} \right| \geqq 0. \tag{23.19}$$

Furthermore

$$\begin{aligned} 0 \leqq I_\nu &= \int_{-\infty}^{\infty} \log \left| \frac{1 - x^2/z_\nu^2}{1 - x^2/\lambda_\nu^2} \right| x^{-2}\, dx \\ &\leqq \frac{1}{\lambda_n} \int_{-\infty}^{\infty} \log \left| \frac{1 - t^2 e^{iO(1/\lambda_n)}}{1 - t^2} \right| t^{-2}\, dt = O\left(\frac{1}{\lambda_n^2}\right). \end{aligned} \tag{23.20}$$

If we integrate term-wise, we obtain

$$0 < \int_{-\infty}^{\infty} \log \left| \frac{\Xi(x)}{H(x)} \right| x^{-2}\, dx = \sum_{\nu=1}^{\infty} I_\nu < \infty. \tag{23.21}$$

Then by (23.17)

$$(\log y)^{-1} \int_{-y}^{y} \log | c^{-1}\, \Xi(x) |\, x^{-2}\, dx \to -\frac{\pi}{2}. \tag{23.22}$$

If we now apply (23.10), and express everything in terms of the zeta function, we obtain

THEOREM XXV.*

$$\int_1^y \log \frac{| \zeta(\tfrac{1}{2} + ix) |}{x^2}\, dx = o(\log y). \tag{23.23}$$

24. **Some theorems of Titchmarsh.** Titchmarsh† has discussed asymptotic properties of entire functions with real negative zeros. In this paragraph we indicate some results which overlap those of Titchmarsh. The method used in deriving these results is closely analogous to that used in proving Theorem XXII; therefore we shall give here only a brief outline of the proof, leaving the details to the reader.

Let

$$f(y) = \prod_{\nu=1}^{\infty} \left(1 + \frac{z}{a_\nu}\right) \tag{24.01}$$

be an entire function all of whose zeros $\{-a_\nu\}$ are negative. It will be assumed that

$$0 < a_1 \leqq a_2 \leqq \cdots, \qquad \sum_{\nu=1}^{\infty} a_\nu^{-1} < \infty. \tag{24.02}$$

* This result is less powerful than one obtained by Titchmarsh in his Cambridge tract on the zeta function. He establishes that

$$\int_0^T \log | \zeta(\tfrac{1}{2} + it) |\, dt = O(T \log \log T).$$

† Loc. cit.

We shall use the symbol $n(r)$ for the number of a_n's not exceeding r. The letter x will designate a real positive variable which tends to infinity.

THEOREM XXVI. *Let λ, ρ, θ be fixed numbers such that*

$$\lambda > 0, \qquad 0 < \rho < 1, \qquad |\theta| < \pi. \tag{24.03}$$

Then the statements

(i) $$n(x) \sim \lambda\, x^{\rho};$$

(ii) $$\log f(x) \sim \pi\,\lambda \operatorname{cosec} \pi\rho\, x^{\rho};$$

(iii) $$\log|f(xe^{i\theta})| \sim \pi\,\lambda \operatorname{cosec} \pi\rho \cos\theta\rho\, x^{\rho} \qquad (|\theta| < \pi);$$

(iv) $$\int_0^x r^{-1-\pi/(2|\theta|)} \log|f(re^{i\theta})|\,dr \sim \frac{\pi\lambda \operatorname{cosec} \pi\rho \cos\theta\rho}{\rho - \dfrac{\pi}{2|\theta|}}\, x^{\rho-\pi/(2|\theta|)} \qquad \left(\frac{\pi}{2\rho} < |\theta| < \pi\right);$$

are all equivalent. In the last statement (iv) *the right-hand member in the case $\rho = \pi/(2\theta)$ should be replaced by its limiting value as $\rho \to \pi/(2\theta)$.*

We first observe that the convergence of the series implies

$$n(x) = o(x). \tag{24.04}$$

Next, let us put

$$\omega(x) = x^{-\rho}\, n(x). \tag{24.05}$$

In view of the fact that $n(x)$ is monotone increasing, it is readily seen that the statements (i), which can be written as

$$\omega(x) \to \lambda, \tag{24.06}$$

and

$$\int_0^x \omega(r)\,dr \sim \lambda x, \tag{24.07}$$

are equivalent.*

Our next step is to transform the left-hand members of (ii–iv) in such a way as to allow an immediate application of Wiener's Tauberian theorems. We have

$$x^{-\rho} \log f(x) = x^{-\rho} \int_0^\infty \log\left(1 + \frac{x}{t}\right) dn(t)$$

$$= x^{-\rho} \int_0^\infty n(t)\, \frac{x}{t(t+x)}\,dt = \frac{1}{x} \int_0^\infty \omega(t)\, \frac{\left(\dfrac{t}{x}\right)^{\rho-1}}{1 + \dfrac{t}{x}}\,dt,$$

* This is readily proved directly or derived from a theorem of Wiener (loc. cit., Theorem XIII, pp. 34–35); it also follows from a well known theorem of Landau, *Beiträge zur analytischen Zahlentheorie*, Rendiconti del Circolo Matematico di Palermo, vol. 26 (1908), pp. 169–302 (p. 218).

$$x^{-\rho}\log|f(xe^{i\theta})| = \frac{1}{2}x^{-\rho}\int_0^\infty \log\left|1+\frac{x}{t}e^{i\theta}\right|^2 dn(t)$$

$$= \frac{1}{x}\int_0^\infty \omega(t)\left(\frac{t}{x}\right)^{\rho-1}\frac{1+\dfrac{t}{x}\cos\theta}{1+2\dfrac{t}{x}\cos\theta+\dfrac{t^2}{x^2}}\,dt,$$

(24.08)

$$x^{\pi/(2|\theta|)-\rho}\int_0^x r^{-1-\pi/(2|\theta|)}\log|f(re^{i\theta})|\,dr$$

$$= x^{\pi/(2|\theta|)-\rho}\int_0^x r^{-1-\pi/(2|\theta|)}\,dr\,r^{\rho-1}\int_0^\infty \omega(t)\left(\frac{t}{r}\right)^{\rho-1}\frac{1+\dfrac{t}{r}\cos\theta}{1+\dfrac{2t}{r}\cos\theta+\dfrac{t^2}{r^2}}\,dt$$

$$= -\frac{1}{x}\int_0^\infty \omega(t)\,dt\left(\frac{x}{t}\right)^{1-\rho+\pi/(2|\theta|)}\int_{x/t}^\infty \frac{1+\dfrac{1}{r}\cos\theta}{1+\dfrac{2}{r}\cos\theta+\dfrac{1}{r^2}}\,r^{-1-\pi/(2|\theta|)}\,dr.$$

For the last transformation use (24.12) with $u = 0$ and $\rho\theta = \pi/2$. Thus all the statements (ii-iv) are expressible in the form

(24.09) $$\frac{1}{x}\int_0^\infty \omega(t)\,N\left(\frac{t}{x}\right)dt \to \lambda \qquad \text{as} \qquad x\to\infty,$$

where $N(y)$ stands, respectively, for

(24.10) $$\begin{cases} N_3(y) = \dfrac{1}{\pi\operatorname{cosec}\pi\rho}\,\dfrac{y^{\rho-1}}{1-y}; \\[2ex] N_4(y) = \dfrac{1}{\pi\operatorname{cosec}\pi\rho\cos\theta\rho}\,y^{\rho-1}\,\dfrac{1+y\cos\theta}{1+2y\cos\theta+y^2}; \\[2ex] N_5(y) = \dfrac{\rho-\dfrac{\pi}{2|\theta|}}{\pi\operatorname{cosec}\pi\rho\cos\theta\rho}\,y^{\rho-1-\pi/(2|\theta|)}\displaystyle\int_{1/y}^\infty \frac{1+\dfrac{1}{r}\cos\theta}{1+\dfrac{2}{r}\cos\theta+\dfrac{1}{r^2}}\,r^{-1-\pi/(2|\theta|)}\,dr. \end{cases}$$

A direct computation yields

(24.11) $$\int_0^\infty \frac{y^{iu+\rho-1}}{1+y}\,dy = \pi\operatorname{cosec}\pi(iu+\rho);$$

(24.12)
$$\int_0^\infty \frac{y^{iu+\rho-1}(1+y\cos\theta)}{1+2y\cos\theta+y^2}\,dy = \frac{1}{2}\left[\int_0^\infty \frac{y^{iu+\rho-1}}{1+y\,e^{i\theta}}\,dy + \int_0^\infty \frac{y^{iu+\rho-1}}{1+y\,e^{-i\theta}}\,dy\right]$$

$$= \int_0^\infty \frac{y^{iu+\rho-1}}{1+y}\,\frac{1}{2}\left[(e^{i\theta})^{-(iu+\rho)}+(e^{-i\theta})^{-(iu+\rho)}\right]dy$$

$$= \pi\operatorname{cosec}\pi(iu+\rho)\cos\theta(iu+\rho);$$

$$\int_0^\infty y^{iu+\rho-1-\pi/(2|\theta|)}\, dy \int_{1/y}^\infty r^{-1-\pi/(2|\theta|)} \frac{1+\dfrac{1}{r}\cos\theta}{1+\dfrac{2}{r}\cos\theta+\dfrac{1}{r^2}}\, dr$$

(24.13)
$$= \int_0^\infty r^{-1+\pi/(2|\theta|)} \frac{1+r\cos\theta}{1+2r\cos\theta+r^2}\, dr \int_r^\infty y^{iu+\rho-1-\pi/(2|\theta|)}\, dy$$

$$= \frac{-\pi \operatorname{cosec} \pi(iu+\rho)\cos\theta(iu+\rho)}{iu+\rho-\dfrac{\pi}{2|\theta|}}.$$

It is an easy matter to verify that the kernels $N_3(y)$, $N_4(y)$ when $|\theta| < \pi/2$, and $N_5(y)$ are possessed of all the properties of the kernel $N(y)$ stated in the proof of Lemma 2.1. We set

$$\Lambda(t) = \int_0^t \omega(t)\, dt.$$

Since $\omega(t) \geqq 0$, $\Lambda(t)$ is monotone increasing. Hence Wiener's theorem used in §21 may be applied here with the result that the statements (ii), (iii) when $|\theta| < \pi/2$ and (iv) are equivalent, while either of (ii) or (iv) implies (iii) when $\pi/2 < |\theta| < \pi$. It should be observed that the kernel $N_4(y)$ is not positive when $|\theta| > \pi/2$ while $N_5(y)$ is positive over the whole range $|\theta| < \pi$. The introduction of this kernel was necessitated by the lack of positiveness of $N_4(y)$ when $|\theta| > \pi/2$. Another theorem of Wiener* will show that either of the statements (ii), (iii) when $|\theta| < \pi/2$ and (iv) implies (24.07), hence (24.06) which is the same as (i). On the other hand, it may be proved directly† that (i) implies (ii), hence also (iii) and (iv). This completes the proof of Theorem XXVI.

25. **A theorem of Pólya.** Pólya‡ has set the problem to establish the following theorem, which we shall number

THEOREM XXVII. *Let $f(z)$ be an entire function bounded for the integral arguments $z = 0, \pm 1, \pm 2, \cdots, \pm n, \cdots$.* Let

(25.01)
$$\max_{0 \leqq \theta < 2\pi} \log |f(re^{i\theta})| = o(r).$$

Then $f(z)$ reduces to a constant.

It is clearly sufficient to prove the theorem for an even $f(z)$, for if $f(z)$ is odd, we need only consider $f(z)/z$, which will be even, and will hence reduce to a con-

* N. Wiener, loc. cit., Theorem XI'', pp. 31–32.

† Titchmarsh, loc. cit., Theorem I.

‡ Jahresbericht der Deutschen Mathematiker Vereinigung, vol. 40 (1931); 2te Abteilung, p. 80, Problem 105. Solutions have been given by Tschakaloff and Szegö, and the solution has been commented on by Pólya (ibid., vol. 43, 2te Abteilung, pp. 10, 11, 67). Pólya refers to an earlier paper of J. M. Whittaker.

stant which can only be zero. The general function may then be treated by reducing it to the sum of an odd and even part.

If $f(z)$ is even,

$$g(z) = [f(z) - f(0)]\, z^{-2} \tag{25.02}$$

will be entire. Thus

$$\sum_{n=-\infty}^{\infty} |g(n)| < \infty . \tag{25.03}$$

Let us form

$$G(z) = \sum_{n=-\infty}^{\infty} \frac{g(n) \sin \pi(n-z)}{\pi(n-z)} . \tag{25.04}$$

Clearly

$$G(x+iy) = O(y^{-1} e^{\pi|y|}) . \tag{25.05}$$

Let us now form the entire function

$$H(z) = [g(z) - G(z)] \operatorname{cosec} \pi z . \tag{25.06}$$

For all values of z and all integral values of n we shall have

$$\begin{aligned} H[(n+\tfrac{1}{2}) + iy] &= O\{\exp \epsilon[(n+\tfrac{1}{2})^2 + y^2]^{1/2}\} + O\left(\frac{1}{|y|}\right) \\ &= e^{\epsilon|n|}\, O(e^{(\epsilon-\pi)|y|}) + O\left(\frac{1}{|y|}\right) \end{aligned} \tag{25.07}$$

uniformly in n. We have here employed (25.01) and (25.05). Hence

$$\int_{-\infty}^{\infty} |H(n+\tfrac{1}{2}+iy)|^2\, dy = O(e^{2\epsilon|n|}) \tag{25.08}$$

for all ϵ.

Let us put

$$x_1 = [x+\tfrac{1}{2}], \qquad x_2 = [x-\tfrac{1}{2}].$$

Then by Cauchy's theorem

$$\begin{aligned} H(x+iy) &= (2\pi i)^{-1} \int_{-\infty}^{\infty} \frac{H(x_1+\frac{1}{2}+iy_1)}{x_1+\frac{1}{2}+iy_1-x-iy}\, dy_1 \\ &\quad - (2\pi i)^{-1} \int_{-\infty}^{\infty} \frac{H(x_2-\frac{1}{2}+iy_1)}{x_2-\frac{1}{2}+iy_1-x-iy}\, dy_1 . \end{aligned} \tag{25.09}$$

Hence

$$\begin{aligned} &\int_{-\infty}^{\infty} H(x+iy)\, e^{iuy}\, dy \\ &= (2\pi i)^{-1} \int_{-\infty}^{\infty} H(x_1+\tfrac{1}{2}+iy_1)\, e^{iuy_1}\, dy_1 \int_{-\infty}^{\infty} \frac{e^{iuy}\, dy}{x_1+\frac{1}{2}-x-iy} \\ &\quad - (2\pi i)^{-1} \int_{-\infty}^{\infty} H(x_2-\tfrac{1}{2}+iy_1)\, e^{iuy_1}\, dy_1 \int_{-\infty}^{\infty} \frac{e^{iuy}\, dy}{x_2-\frac{1}{2}-x-iy} ; \end{aligned} \tag{25.10}$$

and, by the Plancherel theorem and the Schwarz and Minkowski inequalities,

$$\int_{-\infty}^{\infty} |H(x+iy)|^2\,dy$$
$$\leqq \text{const.}\left\{\int_{-\infty}^{\infty} |H(x_1+\tfrac{1}{2}+iy)|^2\,dy + \int_{-\infty}^{\infty} |H(x_2-\tfrac{1}{2}+iy)|^2\,dy\right\}.$$

Thus, by (25.08),

$$\int_{-\infty}^{\infty} |H(x+iy)|^2\,dy = O(e^{2\epsilon|x|}). \tag{25.11}$$

By an application of Cauchy's theorem,

$$\int_{-\infty}^{\infty} H(x+iy)\,e^{iuy}\,dy = e^{-ux}\int_{-\infty}^{\infty} H(iy)\,e^{iuy}\,dy \equiv e^{-ux}\phi(u). \tag{25.12}$$

Thus by the Plancherel Theorem,

$$\int_{-\infty}^{\infty} |\phi(u)|^2\,e^{-2ux}\,du = O(e^{2\epsilon|x|}). \tag{25.13}$$

This is however only possible if $\phi(u)$ vanishes almost everywhere for $|u| > \epsilon$. Since ϵ is arbitrarily small, $\phi(u)$ must be equivalent to zero. Thus $H(z)$ vanishes, and $g(z) = G(z)$. On the other hand, $G(z)$ belongs to L_2 along the real axis since

$$|G(x)|^2 \leqq \sum_{n=-\infty}^{\infty} |ng(n)|^2 \sum_{-\infty}^{\infty} \frac{\sin^2\pi(n-x)}{n^2\pi^2(n-x)^2}; \tag{25.14}$$

$$\int_{-\infty}^{\infty} |G(x)|^2\,dx < \infty. \tag{25.145}$$

Therefore $g(x)$ belongs to L_2 along the real axis. By Theorem XI it must vanish identically. Thus

$$G(z) = g(z) = 0, \qquad f(z) = f(0), \tag{25.15}$$

which is the desired result.

26. Another theorem of Pólya. Pólya* has put the following theorem as a problem, and it has been answered by Szász:†

Let the real numbers $m_1, m_2, \cdots$ *have the properties* $0 < m_1 < m_2 < \cdots$ *and*

$$\lim_{n\to\infty} \frac{n}{m_n} > \frac{b-a}{2\pi} > 0. \tag{26.01}$$

Furthermore, let $f(x)$ *be continuous in the closed interval* (a, b). *Then it will follow from*

$$\int_a^b f(x)\cos m_n x\,dx = \int_a^b f(x)\sin m_n x\,dx = 0 \tag{26.02}$$

that $f(x)$ *vanishes identically.*

* Ibid., vol. 40 (1931), Problem 108 (p. 81).
† Ibid., 43 (1933), p. 20 (part 2).

There is no restriction in supposing $b = -a = \pi$. We shall prove the following more general theorem:

THEOREM XXVIII.* *Let* $0 < m_1 < m_2 < \cdots$ *and let*

$$\overline{\lim_{n\to\infty}} \frac{n}{m_n} > 1. \tag{26.03}$$

Then if $f(x)$ *belongs to* L_2 *and*

$$\int_{-\pi}^{\pi} f(x)\, e^{\pm i m_n x}\, dx = 0 \qquad (n = 1, 2, 3, \cdots), \tag{26.04}$$

$f(x)$ *vanishes except over a set of measure zero.*

It is very important that we have replaced $\underline{\lim}$ by $\overline{\lim}$. This yields us a much deeper theorem.

Since (26.04) is satisfied with $f(x)$ replaced by $f(x) \pm f(-x)$, it suffices to consider the cases of $f(x)$ even or odd. We shall give the discussion of the case $f(x)$ even, under the additional assumption that

$$\int_{-\pi}^{\pi} f(t)\, dt \neq 0.$$

The case where this assumption is not satisfied as well as the case of $f(x)$ odd will require but slight modifications which may be left to the reader. We set

$$\phi(u) = \int_{-\pi}^{\pi} f(t)\, e^{iut}\, dt, \tag{26.05}$$

where the entire function $\phi(u)$ is even, and where we may assume without loss of generality that $\phi(0) = 1$. We observe that, on setting $u = \sigma + i\tau$, we have

$$|\phi(u)| = |\phi(\sigma + i\tau)| \leqq \left\{\int_{-\pi}^{\pi} |f(t)|^2\, dt\right\}^{1/2} e^{\pi|\tau|} = O(e^{\pi|\tau|}). \tag{26.06}$$

On the other hand, we know by the theory of Fourier transforms that $\phi(\sigma)$ belongs to L_2 over $(-\infty, \infty)$, and by (26.05) that the Fourier transform of $\phi(\sigma)$ vanishes outside $(-\pi, \pi)$. Hence by Theorem XII

$$\int_{-\infty}^{\infty} \frac{|\log|\phi(\sigma)||}{1+\sigma^2}\, d\sigma < \infty. \tag{26.07}$$

Thus $\phi(u)$ satisfies the conditions of Theorem XXI. It follows at once that the limit

$$\lim_{r\to\infty} \frac{n_\phi(r)}{r} = A \tag{26.08}$$

* This theorem may also be deduced from a theorem of Titchmarsh: *The zeros of certain integral functions*, Proceedings of the London Mathematical Society, (2), vol. 25 (1925), pp. 283–302, Theorem iv.

exists, where $n_\phi(r)$ is the number of zeros of $\phi(u)$ with modulus not exceeding r. Let $\{u_\nu\}$ be the sequence of the zeros of $\phi(u)$. It is clear that $\{\pm\, m_\nu\}$ is a sub-sequence of $\{u_\nu\}$. Hence by (26.03) we obtain

$$2 < \overline{\lim_{n\to\infty}} \frac{2n}{m_n} \leqq \lim_{r\to\infty} \frac{n_\phi(r)}{r} = A\,. \tag{26.09}$$

However, by Jensen's theorem, in view of (26.06),

$$\begin{aligned} \frac{1}{r}\int_0^r \frac{n_\phi(t)}{t}\,dt &= \frac{1}{2\pi r}\int_0^{2\pi} \log|\phi(re^{i\theta})|\,d\theta \\ &\leqq \frac{1}{2\pi r}\int_0^{2\pi} \pi r\,|\sin\theta|\,d\theta + O\left(\frac{1}{r}\right) = 2 + O\left(\frac{1}{r}\right). \end{aligned} \tag{26.10}$$

Hence

$$A = \lim_{r\to\infty} \frac{1}{r}\int_0^r \frac{n_\phi(t)}{t}\,dt \leqq 2\,. \tag{26.11}$$

The resulting contradiction shows that $f(x)$ must vanish except for a set of measure zero.

CHAPTER VI

THE CLOSURE OF SETS OF COMPLEX EXPONENTIAL FUNCTIONS

27. Methods from the theory of entire functions. The chief object of this chapter is to discuss the closure of the set of functions $\{e^{\pm i\lambda_n x}, 1\}$ over a finite interval, which we may take without restriction to be $(-\pi, \pi)$. This is a subject with a surprisingly small literature. The entire literature seems to center around a paper of Birkhoff* in which he uses a method of continuous variation to handle the problem of the closure of sets of Sturm-Liouville functions. The ideas of Birkhoff's theorem have been applied to trigonometric sets of functions by Walsh.† As far as we know, the only discussion of a case where the sole restriction on λ_n (besides reality and evenness) is of the form

$$|\lambda_n - n| < L < \infty \tag{27.01}$$

is due to Wiener.‡ The present chapter, and part of the next one, will be devoted to results similar to Wiener's, but of greater scope.

In §26, we have shown that if

$$\varlimsup_{n\to\infty} \frac{\lambda_n}{n} < 1, \tag{27.02}$$

the set of functions $\{e^{\pm i\lambda_n x}\}$ is closed L_2 over$(-\pi, \pi)$. In this section, we shall confine ourselves to the case where

$$\lim_{n\to\infty} \frac{\lambda_n}{n} = 1. \tag{27.03}$$

In this case, the entire function

$$F(z) = \prod_1^\infty \left(1 - \frac{z^2}{\pi^2 \lambda_n^2}\right) \tag{27.04}$$

will exist. If $\Lambda(t)$ is the number of λ_n's less than t,

$$\begin{aligned} \frac{1}{y}\log F(iy) &= \frac{1}{y}\int_0^\infty \log\left(1 + \frac{y^2}{\pi^2 t^2}\right) d\Lambda(t) \\ &= \int_0^\infty \frac{\Lambda(t)}{t}\,\frac{2y}{\pi^2 t^2 + y^2}\,dt. \end{aligned} \tag{27.045}$$

* G. D. Birkhoff, *A theorem on series of orthogonal functions with an application to Sturm-Liouville series*, Proceedings of the National Academy of Sciences, vol. 3 (1917), p. 656.

† J. L. Walsh, *A generalization of the Fourier cosine series*, Transactions of the American Mathematical Society, vol. 22 (1921), pp. 230–239.

‡ N. Wiener, *On the closure of certain assemblages of trigonometrical functions*, Proceedings of the National Academy of Sciences, vol. 13 (1927), p. 27.

If we now use (27.03), we have

$$\lim_{y\to\infty} \frac{1}{y} \log F(iy) = 1. \tag{27.05}$$

Throughout this chapter, we shall define $F(z)$ as in (27.04). Assuming (27.03) to be true, we shall prove a series of theorems connecting the closure properties of $\{e^{\pm i\lambda_n x}\}$ or of $\{1, e^{\pm i\lambda_n x}\}$ with the properties of the numbers λ_n or of the function $F(z)$. Of these the first is

THEOREM XXIX. *Let (27.03) be true and let $F(z)$ belong to L_2 over the real axis. Then the set of functions $\{e^{\pm i\lambda_n x}\}$ cannot be closed L_2 over $(-\pi, \pi)$. Again, let $zF(z)$ belong to L_2 over the real axis. Then the set of functions $\{1, e^{\pm i\lambda_n x}\}$ cannot be closed L_2 over $(-\pi, \pi)$. In either case, a finite number of the functions of the set may be replaced by other functions of the form $e^{i\lambda x}$ to the same number.*

If $F(z)$ belongs to L_2, let it have the Fourier transform

$$f(u) = \underset{A\to\infty}{\text{l.i.m.}} \frac{1}{(2\pi)^{1/2}} \int_{-A}^{A} F(x)\, e^{iux}\, dx, \tag{27.06}$$

and let us put

$$H(z) = \int_0^z \frac{F(w)}{w+i}\, e^{iw(1+\epsilon)}\, dw, \qquad \epsilon > 0. \tag{27.07}$$

Then $H(z)$ will be an entire function of order 1, which will be bounded both along the real axis and the positive imaginary axis. Thus by the Phragmén-Lindelöf theorem, it will be bounded in the whole upper half-plane $\Im(z) > 0$. By Cauchy's theorem,

$$\frac{F(z)\, e^{iz(1+\epsilon)}}{z+i} = \frac{1}{2\pi i} \int_{|w-z|=\delta} \frac{H(w)}{(w-z)^2}\, dw. \tag{27.08}$$

Thus $F(z)\, e^{iz(1+\epsilon)}/(z+i)$ is bounded in any half-plane above $\Im(z) = \delta > 0$. It may similarly be proved that $F(z)\, e^{-i(z+\epsilon)}/(z-i)$ is bounded in any half-plane below $\Im(z) = -\delta < 0$. Thus by the Phragmén-Lindelöf theorem, $F(z)/(z+i)$ is bounded on every horizontal strip above the line $\Im(z) = -1$. Thus $F(z)\, e^{iz(1+\epsilon)}/(z+i)$ is bounded in *every* upper half-plane $\Im(z) > a > -1$. It follows that for arbitrary ϵ,

$$\begin{aligned} &\lim_{\Im(z)\to\infty} \underset{-\infty<\Re(z)<\infty}{\text{l.u.b.}} \left| \frac{F(z)\, e^{iz(1+\epsilon)}}{z+i} \right| \\ &= \lim_{\Im(z)\to\infty} e^{-\Im(z)\epsilon/2} \underset{-\infty<\Re(z)<\infty}{\text{l.u.b.}} \left| \frac{F(z)\, e^{iz(1+\epsilon/2)}}{z+i} \right| < K \lim_{\Im(z)\to\infty} e^{-\Im(z)\epsilon/2} = 0. \end{aligned} \tag{27.09}$$

Now let us apply Cauchy's theorem. If A and B are sufficiently large, we have

$$\begin{aligned} \frac{F(z)\, e^{zi(1+\epsilon)}}{(z+i)^2} = \frac{1}{2\pi i} &\left[\int_{-A}^{A} + \int_{A}^{A+iB} + \int_{A+iB}^{-A+iB} \right. \\ &\left. + \int_{-A+iB}^{-A} \right] \frac{F(w)\, e^{iw(1+\epsilon)}}{(w+i)^2\,(w-z)}\, dw. \end{aligned} \tag{27.10}$$

Since F is bounded, if we let A tend to infinity, we get

$$\text{(27.11)} \qquad \frac{F(z)\,e^{iz(1+\epsilon)}}{(z+i)^2} = \frac{1}{2\pi i}\left[\int_{-\infty}^{\infty} - \int_{-\infty+iB}^{\infty+iB}\right] \frac{F(w)\,e^{iw(1+\epsilon)}}{(w+i)\,(w-z)}\,dw.$$

Now let B tend to infinity. By (27.09) we have

$$\text{(27.12)} \qquad \frac{F(z)\,e^{iz(1+\epsilon)}}{(z+i)^2} = \frac{1}{2\pi i}\int_{-\infty}^{\infty} \frac{F(w)\,e^{iw(1+\epsilon)}}{(w+i)^2\,(w-z)}\,dw.$$

By the Parseval theorem, we have

$$\text{(27.13)} \qquad \begin{aligned} &= \operatorname*{l.i.m.}_{A\to\infty} \frac{1}{(2\pi)^{1/2}} \int_{-A+yi}^{A+yi} \frac{F(w)\,e^{iw(1+\epsilon)}}{(w+i)^2}\,e^{iwx}\,dw \\ &= \begin{cases} e^{yx} \operatorname*{l.i.m.}_{A\to\infty} \displaystyle\int_{-A}^{A} F(w)\,e^{iw(1+\epsilon)}\,e^{iwx}\,dw & [x<0], \\ 0 \qquad [x>0]. \end{cases} \end{aligned}$$

From this it results that

$$\text{(27.14)} \qquad \lim_{y\to 0} \int_{-\infty}^{\infty} dw \left| \frac{F(w+iy)\,e^{i(w+iy)(1+\epsilon)}}{(w+iy+i)^2} - \frac{F(w)\,e^{iw(i+\epsilon)}}{(w+i)^2} \right|^2 = 0.$$

Here we use the fact that, by (27.13), $F(w+iy)e^{i(w+iy)(i+\epsilon)}/(w+iy+1)^2$ converges in the mean as $y \to 0$, and hence must converge to its ordinary limit. Thus by the Plancherel theorem,

$$\text{(27.15)} \qquad \operatorname*{l.i.m.}_{A\to\infty} \frac{e^{-x}}{(2\pi)^{1/2}} \int_{-A}^{A} \frac{F(w)\,e^{iw(1+\epsilon)}}{(w+i)^2}\,e^{iwx}\,dw = 0 \qquad [x>0].$$

Hence by two formal differentiations,

$$\text{(27.16)} \qquad \operatorname*{l.i.m.}_{A\to\infty} \frac{1}{(2\pi)^{1/2}} \int_{-A}^{A} F(w)\,e^{iw(1+\epsilon)}\,e^{iwx}\,dw = 0 \qquad [x>0].$$

These formal differentiations may be justified by the fact that $F(z)$ belongs to L_2, and that

$$\text{(27.17)} \qquad \begin{aligned} &\int_0^x \left\{ \operatorname*{l.i.m.}_{A\to\infty} \frac{e^{-x}}{(2\pi)^{1/2}} \int_{-A}^{A} F(z)\,e^{iz(1+\epsilon)}\,e^{izx}\,dz \right\} dx \\ &= \frac{e^{-x}}{i(2\pi)^{1/2}} \int_{-\infty}^{\infty} \frac{F(z)\,e^{iz(1+\epsilon)}}{z+i}\,e^{izx}\,dz - \frac{1}{i(2\pi)^{1/2}} \int_{-\infty}^{\infty} \frac{F(z)\,e^{iz(1+\epsilon)}}{z+i}\,dz. \end{aligned}$$

If we discard the constant term and integrate once again, we see that we have (27.15).

Similarly,

$$\text{(27.18)} \qquad \operatorname*{l.i.m.}_{A\to\infty} \frac{1}{(2\pi)^{1/2}} \int_{-A}^{A} F(z)\,e^{-iz(1+\epsilon)}\,e^{izx}\,dz = 0 \qquad [x<0].$$

Thus $f(u/\pi)$, defined in (27.06), is a function in L_2, for which

$$f(u/\pi) = 0 \qquad [|u| > \pi], \tag{27.19}$$

and

$$\int_{-\infty}^{\infty} f\left(\frac{u}{\pi}\right) e^{\pm iu\lambda_n}\, du = \pi(2\pi)^{1/2} F(\pm \pi\lambda_n) = 0 \qquad [n = 1, 2, \cdots]. \tag{27.20}$$

Hence the set of functions $\{e^{\pm i\lambda_n u}\}$ cannot be closed over $(-\pi, \pi)$.

The part of Theorem XXIX which refers to $\{1, e^{\pm i\lambda_n u}\}$ is proved in exactly the same manner.

The possibility of replacing a finite number of functions by other functions of similar form results from the fact that the replacement of a finite number of linear factors of $F(z)$ by an equal number of other linear factors affects neither the question as to whether $F(z)$ belongs to L_2 along the real axis nor as to the range of values over which its Fourier transform differs from zero. The second of these statements results from the fact that

$$\lim_{y\to\infty} \frac{1}{y} \log F(iy) \tag{27.21}$$

is not affected.

We now proceed to

Theorem XXX. *Let (27.03) be true and let the set of functions $\{e^{\pm i\lambda_n x}\}$ be closed L_2 over $(-\pi, \pi)$ but let it cease to be closed on the removal of some one term. Then it ceases to be closed on the removal of any one term, $F(z)$ does not belong to L_2 along the real axis, but $F(z)/z$ belongs to L_2 over $(1, \infty)$. Again, if the set of functions $\{1, e^{\pm i\lambda_n x}\}$ is closed L_2 over $(-\pi, \pi)$, but ceases to be closed on the removal of some one term, this term is arbitrary, $zF(z)$ does not belong to L_2 along the real axis but $F(z)$ belongs to L_2.*

We only need consider the first part of this theorem. It follows at once from Theorem XXIX that if the hypothesis of this theorem is satisfied, $F(z)$ does not belong to L_2. On the other hand, there is some non-null function of L_2, say $\phi(x)$, which is orthogonal over $(-\pi, \pi)$ to all but one of the functions $\{e^{\pm i\lambda_n x}\}$. Let us put

$$\Phi(z) = \int_{-\pi}^{\pi} \phi(x)\, e^{ixz}\, dx\,. \tag{27.22}$$

Since by the Schwarz inequality

$$\Phi(z) = O(e^{\pi|z|}), \tag{27.23}$$

we see that $\Phi(z)$ is an entire function of at most order 1, with zeros at $z = \mu_n$. The numbers $\{\mu_n\}$ contain all but one, say μ, of the set $\{\lambda_n\}$. By known properties of entire functions,

$$\sum_1^{\infty} \frac{1}{|\mu_n|^{1+\epsilon}} < \infty\,. \tag{27.24}$$

Thus we have

$$(27.25)\qquad \Phi(z) = A\left\{\prod_1^\infty\left(1-\frac{z}{\mu_n}\right)e^{z/\mu_n}\,e^{cz}\right\}.$$

Thus there is an entire function, say $\Psi(z)$, such that

$$(27.26)\qquad \Phi(z) = \frac{F(\pi z)}{z-\mu}\,\Psi(z).$$

If $\Psi(z)$ is not an exponential, it will have at least one root, which we may write v. Then as $|z| \to \infty$,

$$(27.27)\qquad \Phi(z) \sim F(\pi z)\,\frac{\Psi(z)}{z-v}.$$

Thus $F(\pi u)\,\Psi(u)/(u - v)$ will belong to L_2 over the entire real axis of u, for $\Phi(z)$ belongs to L_2, by (27.22) and the Plancherel theorem. Let

$$(27.28)\qquad g(x) = \underset{A\to\infty}{\text{l.i.m.}}\,\frac{1}{2\pi}\int_{-A}^{A} F(\pi u)\,\Psi(u)\,e^{-iux}\,du.$$

It will then follow by the argument of Theorem XXVII (cf. (27.19)) that

$$(27.29)\qquad g(x) = 0 \qquad [\,|x| > \pi\,].$$

Thus by the Plancherel theorem,

$$(27.30)\qquad \int_{-\pi}^{\pi} g(x)\,e^{\pm i\lambda_n x}\,dx = \frac{F(\pm\,\pi\lambda_n)}{\pm\,\lambda_n - v}\,\Psi(\pm\lambda_n) = 0,$$

while $g(x)$ belongs to L_2. This contradicts our assumption of the closure of $\{e^{\pm i\lambda_n x}\}$ and hence $\Psi(z)$ has no zeros. From (27.25) we have therefore

$$(27.31)\qquad \Phi(z) = \frac{AF(\pi z)\,e^{cz}}{z-\mu}.$$

Since on the real axis $\Phi(z)$ belongs to L_2, and since $F(\pi z)$ is even, we see that $F(\pi z)/(z - \mu)$ must be of class L_2 on the real axis. Thus our theorem is established.

We now come to

Theorem XXXI. *Let (27.03) be true and let*

$$(27.32)\qquad |F(x+i\epsilon)| \geqq \frac{A}{1+|x|^n} > 0$$

for all real x and some n and ϵ. Then the set of functions $\{e^{\pm i\lambda_n x}\}$ will be closed L_2 or not closed L_2 over $(-\pi, \pi)$, according as $F(z)$ does not or does belong to L_2 over the entire real axis. It can always be made closed by the adjunction of a finite number of functions $e^{i\lambda x}$. The set of functions $\{1, e^{\pm i\lambda_n x}\}$ will be closed or not closed L_2 according as $zF(z)$ does not or does belong to L_2 over the entire real axis.

This theorem also depends on the use of a function $\phi(x)$ and its Fourier transform $\Phi(z)$. Let $\{e^{\pm i\lambda_n x}\}$ be closed. Then by Theorem XXIX $F(z)$ does not belong to L_2. On the other hand, let $\{e^{\pm i\lambda_n x}\}$ not be closed, and let

$$(27.33) \qquad \int_{-\pi}^{\pi} \phi(x)\, e^{\pm i\lambda_n x}\, dx = 0 \qquad (n = 1, 2, 3, \cdots).$$

Let $\Phi(z)$ be defined as in (27.22). As in the proof of Theorem XXX,

$$(27.34) \qquad \Phi(z) = F(\pi z)\, \Psi(z),$$

where $\Psi(z)$ is an entire function. It is clear from our asymptotic estimate of $F(z)$ along the imaginary axis and from the fact that

$$(27.35) \qquad \begin{aligned} |\Phi(iy)| = \left| \int_{-\pi}^{\pi} \phi(x)\, e^{-xy}\, dy \right| &\leqq \left\{ \int_{-\pi}^{\pi} |\phi(x)|^2\, dx \right\}^{1/2} \left\{ \int_{-\pi}^{\pi} e^{-2xy}\, dx \right\}^{1/2} \\ &= O(e^{\pi |y|}), \end{aligned}$$

that for all $\epsilon_1 > 0$,

$$(27.36) \qquad \Psi(iy) = O(e^{\epsilon_1 |y|}).$$

Again, along the line $\mathfrak{I}(z) = \epsilon$, $\Phi(z)$ is bounded, and by (27.32)

$$(27.37) \qquad \Psi(z) \leqq B(1 + |z|^n).$$

The function $\Psi(z)$ is an entire function of order 1 or less, and a simple application of the Phragmén-Lindelöf theorem will yield us as a corollary of (27.36) and (27.37),

$$(27.38) \qquad \Psi(z) = O(e^{\epsilon_1 |z|}).$$

Now let

$$(27.39) \qquad \Psi(z) = (z - \alpha_1)(z - \alpha_2) \cdots (z - \alpha_k)\, \Psi_1(z),$$

where $\Psi_1(z)$ is an entire function. This will always be possible unless $\Psi(z)$ is a polynomial. The entire function $\Psi_1(z)$ will belong to L_2 along some parallel to the real axis, and will satisfy

$$(27.40) \qquad \Psi_1(z) = O(e^{\epsilon_1 |z|}).$$

Thus by an argument essentially the same as that which we have used in proving Theorem XXIX, the Fourier transform of $\Psi_1(z)$ along the straight line along which we have shown it to belong to L_2 will differ from 0 only outside $(-\epsilon_1, \epsilon_1)$. Since ϵ_1 is arbitrarily small, this Fourier transform and hence $\Psi_1(z)$ itself must be equivalent to zero. Thus $\Phi(z)$ will also have to vanish identically, which gives us a contradiction. This shows that $\Psi(z)$ must be a polynomial. From (27.34) this shows that $F(\pi z)$ belongs to L_2 along the real axis since $\Phi(z)$ does. To show it can be closed by the addition of a finite number of terms $e^{i\lambda x}$, we adjoin enough terms of this form so that only one is lacking for closure. Then we construct $\phi(x)$ orthogonal to all the terms. Proceeding as in the proof above, it is readily seen that only a finite number of terms have been added.

The remainder of Theorem XXXI is trivial, or is proved in the same way as the first part. One step which is of some interest asserts that either both or neither of the functions $F(x)x^m$ and $F(x + i\epsilon)x^m$ belong to L_2. This may be proved by applying the argument of Theorem XXIX to $F(x)x^m$ instead of to $F(x)$.

A direct corollary of Theorems XXIX and XXXI is

THEOREM XXXII. *Let* (27.32) *hold true for some* n *and* ϵ *and let*

$$|F(x)| \leqq \text{const. } (1 + |x|^m) \tag{27.41}$$

for real x. *Let us call a set of functions exact over* $(-\pi, \pi)$ *when it is closed, but ceases to be closed on the removal of any one term. Then the set of functions* $\{e^{\pm i\lambda_n x}\}$ *is either exact as it stands, or becomes exact on the removal of a finite number of functions, or becomes exact on the adjunction of a finite number of functions* $e^{i\lambda x}$. *The number of functions removed or adjoined will be called the excess or deficiency, respectively, of the set. The particular functions of the form* $e^{i\lambda x}$ *removed or adjoined are arbitrary; only their number is significant.*

An important application of Theorem XXXII is the proof of

THEOREM XXXIII. *If* (27.01) *holds,* $\{e^{\pm i\lambda_n x}\}$ *is of at most finite excess or deficiency.*

The proof of this theorem depends on an estimate of

$$F(\pi w) = \prod_1^\infty \left(1 - \frac{w^2}{\lambda_n^2}\right) \tag{27.42}$$

when (27.01) is fulfilled. If $|\mathfrak{I}(w)| = \epsilon > 0$ and if $|\mathfrak{R}(w)| = u > 2L$, we have

$$\begin{aligned} |F(\pi w)| &= \left|\prod_1^\infty \left(1 - \frac{w^2}{\lambda_n^2}\right)\right| \\ &= \prod_1^{[|u|]-[L]-1} \left|1 - \frac{w^2}{\lambda_n^2}\right| \prod_{[|u|]-[L]}^{[|u|]+[L]+1} \left|1 - \frac{w^2}{\lambda_n^2}\right| \prod_{[|u|]+[L]+2}^{\infty} \left|1 - \frac{w^2}{\lambda_n^2}\right|. \end{aligned} \tag{27.43}$$

For the terms in the first product of the right hand member we have

$$\left|1 - \frac{w^2}{\lambda_n^2}\right| = \left|1 - \frac{(u \pm i\epsilon)^2}{\lambda_n^2}\right| \geqq \left|1 - \frac{w^2}{(n + L)^2}\right|. \tag{27.44}$$

For the terms in the second product we have

$$\left|1 - \frac{w^2}{\lambda_n^2}\right| = \left|\frac{\lambda_n^2 - u^2 \pm 2i\epsilon u + \epsilon^2}{\lambda_n^2}\right| \geqq \frac{2\epsilon|u|}{\lambda_n^2} \geqq \frac{\epsilon}{2u}. \tag{27.45}$$

For the terms in the third product we have

$$(27.46)\qquad \left|1-\frac{w^2}{\lambda_n^2}\right| \geqq \left|1-\frac{w^2}{(n-L)^2}\right|.$$

Thus

$$(27.47)\qquad \begin{aligned} |F(\pi w)| &\geqq \left(\frac{\epsilon}{2}\right)^{2[L]+2} |w|^{-2[L]-2} \\ &\times \prod_{1}^{[|u|]-[L]-1} \left|1-\frac{w^2}{(n+L)^2}\right| \prod_{[|u|]+[L]+2}^{\infty} \left|1-\frac{w^2}{(n-L)^2}\right|. \end{aligned}$$

But

$$(27.48)\qquad \prod_{1}^{[|u|]-[L]-1} \left|1-\frac{w^2}{(n+L)^2}\right| \geqq \prod_{[L]+2}^{[|u|]+1} \left|1-\frac{w^2}{n^2}\right|,$$

and

$$(27.49)\qquad \prod_{[|u|]+[L]+2}^{\infty} \left|1-\frac{w^2}{(n-L)^2}\right| \geqq \prod_{[|u|]+1}^{\infty} \left|1-\frac{w^2}{n^2}\right|.$$

Therefore

$$(27.50)\qquad \begin{aligned} |F(\pi w)| &\geqq \left(\frac{\epsilon}{2}\right)^{2[L]+2} |w|^{-2[L]-2} \prod_{[L]+2}^{\infty} \left|1-\frac{w^2}{n^2}\right| \\ &= \frac{1}{\pi}\left(\frac{\epsilon}{4}\right)^{2[L]+2} |W|^{-4[L]-5}\, \pi w \prod_{1}^{\infty} \left|1-\frac{w^2}{n^2}\right| \\ &= \frac{1}{\pi}\left(\frac{\epsilon}{4}\right)^{2[L]+2} |w|^{-4[L]-5} \sin \pi w \\ &\geqq \left(\frac{\epsilon}{4}\right)^{2[L]+3} |w|^{-4[L]-5}. \end{aligned}$$

Under the same circumstances

$$(27.51)\qquad \begin{aligned} |F(\pi w)| &= \prod_{1}^{[L]+1} \left|1-\frac{w^2}{\lambda_n^2}\right| \prod_{[L]+2}^{[|u|]} \left|1-\frac{w^2}{\lambda_n^2}\right| \prod_{[|u|]+1}^{\infty} \left|1-\frac{w^2}{\lambda_n^2}\right| \\ &\leqq A\, |w|^{2[L]+2} \prod_{1}^{[|u|]-[L]-1} \left|1-\frac{w^2}{n^2}\right| \prod_{[|u|]+[L]+2}^{\infty} \left|1-\frac{w^2}{n^2}\right|. \end{aligned}$$

But

$$(27.52)\qquad \prod_{[|u|]-[L]}^{[|u|]+[L]+1} \left|1-\frac{w^2}{n^2}\right| \geqq \prod_{[|u|]-[L]}^{[|u|]+[L]+1} \frac{\epsilon\,|u|}{n^2} \geqq |6w|^{-2[L]-2}.$$

Thus

$$|F(\pi w)| \leqq A |w|^{4[L]+3} \sin \pi w \leqq A |w|^{4[L]+3}. \tag{27.53}$$

Here A is a constant which may be different in each expression. We may now apply the Phragmén-Lindelöf theorem to obtain the fact that if x tends to infinity through real values,

$$|F(\pi x)| \leqq \text{const.} |x|^{4[L]+3}. \tag{27.54}$$

If we now apply Theorem XXXII, Theorem XXXIII follows at once.

A specialization of Theorem XXXIII is

THEOREM XXXIV. *Let* (27.01) *hold, and let*

$$L < \frac{n}{4} + \frac{1}{8}. \tag{27.55}$$

Then the deficiency of $\{1, e^{\pm i\lambda_n x}\}$ *cannot exceed* n. *If*

$$L \leqq \frac{n}{4} + \frac{1}{8} \tag{27.56}$$

then the excess of $\{1, e^{\pm i\lambda_n x}\}$ *cannot exceed* n.

Theorem XXXIV differs from Theorem XXXIII in that we have to make use of more refined estimates depending on the gamma function. These all go back to Stirling's formula. In particular, if x is real and B is an integer, we have

$$\prod_1^\infty \left|1 - \frac{(x + i\epsilon)^2}{(n + B)^2}\right| \geqq K_1 |x|^{-1-2B} \tag{27.57}$$

and

$$\prod_1^\infty \left|1 - \frac{x^2}{(n + B)^2}\right| \leqq K_2 |x|^{-1-2B}. \tag{27.58}$$

In our proof we shall use the notation

$$0 < C_1 \leqq C \leqq C_2 < \infty, \tag{27.59}$$

where C may be variable, but C_1 and C_2 are fixed. We shall take ϵ fixed, and shall put

$$v = [\Re(w)]; \qquad |\Im(w)| = \epsilon: \tag{27.60}$$

Then if B is an integer and $A + B \geqq \Re(w) - v$, A and B being taken as constants,

$$\prod_{1}^{v+B}\left|1-\frac{w^2}{(n+A)^2}\right|$$

(27.61)
$$= \frac{|\Gamma(v+A+B-w+1)\,\Gamma(v+A+B+w+1)|\,\{\Gamma(A+1)\}^2}{\{\Gamma(v+A+B+1)\}^2\,|\Gamma(A-w+1)\,\Gamma(A+w+1)|}$$
$$= \frac{c\,|\Gamma(v+A+B+w+1)\,\Gamma(w-A)|}{\{\Gamma(v+A+B+1)\}^2\,|\Gamma(A+w+1)|}$$
$$= c\,2^{2\Re(w)}\,|w|^{\Re(w)-v-3A-B-3/2}.$$

Hence if the imaginary part of w is ϵ, and the real part is sufficiently large and positive

(27.62)
$$|F(\pi w)| \geqq a \prod_{1}^{v-[L]-1}\left|1-\frac{w^2}{(n+L)^2}\right| |w|^{-2[L]-2} \prod_{v+[L]+2}^{\infty}\left|1-\frac{w^2}{(n-L)^2}\right|$$
$$\geqq a\,2^{2\Re(w)}\,|w|^{-3L+[L]-1/2+\Re(w)-v}\,|w|^{-2[L]-2}\prod_{v+2}^{\infty}\left|1-\frac{w^2}{n^2}\right|$$
$$\geqq a\,|w|^{-3L+[L]-1/2+\Re(w)-v}\,|w|^{-2[L]-2}\,|w|^{-\Re(w)+v+5/2}\prod_{1}^{\infty}\left|1-\frac{w^2}{n^2}\right|$$
$$\geqq a\,|w|^{-4L-1},$$

where a is a positive constant which may be different in each expression. Again,

(27.63)
$$|F(\pi w)| \leqq \prod_{1}^{[L]+1}\left|1-\frac{w^2}{\lambda_n^2}\right| \prod_{[L]+2}^{v}\left|1-\frac{w^2}{(n-L)^2}\right| \prod_{v+1}^{\infty}\left|1-\frac{w^2}{(n+L)^2}\right|$$
$$\leqq a\,|w|^{2[L]+2}\prod_{2}^{v-[L]}\left|1-\frac{w^2}{n^2}\right| \prod_{v+[L]+1}^{\infty}\left|1-\frac{w^2}{n^2}\right|$$
$$\leqq a\,|w|^{2[L]-1}\,|\sin \pi w|\,|w|^{2[L]} \leqq a\,|w|^{4L-1}.$$

Hence as in the proof of Theorem XXXIII when x is real and sufficiently large,

(27.64)
$$|F(\pi x)| \leqq \text{const.}\,|x|^{4L-1}.$$

Thus $F(x)/(|x|^n+1)$ belongs to L_2 along the real axis, when $L < n/4 + 1/8$. Again $F(w)w^{1+n}$ fails to belong to L_2 along a parallel to the real axis when $L \leqq n/4 + 1/8$. It follows immediately from the discussion of Theorem XXIX that this will be true when and only when $F(x)x^{1+n}$ fails to belong to L_2 along the real axis. Thus Theorem XXXIV which restricts the excess and the deficiency of $\{1, e^{\pm i\lambda_n x}\}$ follows at once from Theorem XXXI.

28. **The duality between closure and independence.** Let the set of functions $\{f_n(x)\}$ be closed, normal, and orthogonal over (a, b). Let $a < c < b$, and let the set $\{f_n(x)\}$ be made up of the two non-overlapping sub-sets $\{g_n(x)\}$ and

$\{h_n(x)\}$. Let $f(x)$ vanish over (a, c), and let it belong to L_2 over (c, b). Then if $f(x)$ is orthogonal to every function of $\{g_n(x)\}$ over (c, b), it is also orthogonal over (a, b), and has an orthogonal development

$$f(x) \sim \sum_{1}^{\infty} a_n h_n(x) \tag{28.01}$$

for which

$$\sum_{1}^{\infty} |a_n|^2 < \infty . \tag{28.02}$$

It follows that if $\{g_n(x)\}$ is not closed over (c, b), then there is a series (28.01) satisfying (28.02), and converging in the mean to zero over (a, c).

On the other hand, if $\{g_n(x)\}$ is closed over (c, b), there is no function $f(x)$ vanishing over (a, c), belonging to L_2 over (c, b), and there orthogonal to every function $g_n(x)$. It follows that there is no function with an orthogonal development (28.01) for which (28.02) holds, vanishing over (a, c). We shall express this result by writing

THEOREM XXXV. *Let $\{f_n(x)\}$ be a set of functions normal and orthogonal and closed over (a, b). We shall call a sub-set $\{h_n(x)\}$ weakly independent over a sub-interval (a, c) when the relation*

$$0 \sim \sum_{-\infty}^{\infty} a_n h_n(x) \tag{28.03}$$

over (a, c) for which (28.02) is valid implies that $a_n = 0$ for all n. Then $\{h_n(x)\}$ is weakly independent over (a, c) when and only when the complementary set $\{g_n(x)\}$, consisting of all the members of $\{f_n(x)\}$ not belonging to $\{h_n(x)\}$, is closed over (c, b). In particular, a linearly independent set is weakly independent, so that its complementary set is always closed. If a set is both closed and weakly independent, so is its complementary set over the complementary range. If a set and its complementary are both independent over complementary ranges, they are both closed.

As an application, let the set $\{1, e^{\pm inx}\}$ be made up of the sets $\{e^{\pm i\lambda_n x}\}$ and $\{1, e^{\pm i\mu_n x}\}$. Let $\lambda_1 \leqq \lambda_2 \leqq \cdots \leqq \lambda_n \leqq \cdots$ and let

$$\lim_{n \to \infty} \frac{n}{\lambda_n} = 0 . \tag{28.04}$$

By Theorem XXVIII, it will follow that the set of functions $\{e^{\pm i\mu_n x}\}$ is closed over *any* interval $(-\pi + \epsilon, \pi)$ and hence over any interval $(A, A + 2\pi - \epsilon)$, or over any set $(-\pi, A - \epsilon) + (A, \pi)$ for which $-\pi + \epsilon < A < \pi$. By Theorem XXXV, the complementary set of functions $\{e^{\pm i\lambda_n x}\}$ will be weakly independent over any interval $(A - \epsilon, A)$, no matter how small. We thus have established

THEOREM XXXVI. *Let $\lambda_{-n} = -\lambda_n$, $\lambda_1 \leqq \lambda_2 \leqq \cdots \leqq \lambda_n \leqq \cdots$, let all the λ_n's be integers, and let*

(28.05) $$\lim_{n\to\infty} \frac{n}{\lambda_n} = 0 .$$

Let $\sum_{-\infty}^{\infty} | a_n |^2 < \infty$, and let

(28.06) $$\underset{N\to\infty}{\text{l.i.m.}} \sum_{-N}^{N} a_n e^{i\lambda_n x} = 0 \qquad (A - \epsilon \leqq x \leqq A),$$

where ϵ is any positive quantity. Then the coefficients a_n are all identically zero.

A more general result of the same nature is obtained as follows: let λ_n be a set of positive numbers, and let

(28.07) $$\text{g.l.b.} \left| \lambda_m - \lambda_n \right| > L > 0 .$$

There exists a number D such that

(28.08) $$L \geqq D \geqq L(1 - \epsilon), \qquad \epsilon > 0,$$

and such that no $\lambda_n = (n + \frac{1}{2}) D$. To see this, we observe that the numbers L_n for which

(28.09) $$\lambda_p = (m + \tfrac{1}{2}) L_n$$

for some pair of integers p and m are denumerable, since a denumerable set of denumerable sets is denumerable.

Now let μ_n stand for $(n + \frac{3}{4})D$ if there is no λ_m nearer to nD than $D/2$. Let μ_n stand for λ_m if

(28.10) $$nD < \lambda_m \leqq (n + 1) D \qquad (n = 0, 1, \cdots).$$

Let $\sigma_{2n} = \mu_n$ and $\sigma_{2n+1} = (2n + 1)D - \mu_n$. Then there can be no $\sigma_n = \sigma_m$, $m \neq n$. We shall now show that the set of functions $\{1, e^{\pm i\sigma_n x}\}$ is closed over $(-2\pi/D, 2\pi/D)$, and ceases to be closed if we remove any one term.

We form

(28.11) $$f(z) = z \prod_{1}^{\infty} \left(1 - \frac{z^2}{\sigma_n^2}\right),$$

and consider it on the line $\Im(z) = \delta > 0$. We have

(28.12) $$\frac{\left(1 - \dfrac{z^2}{\sigma_{2n}^2}\right)\left(1 - \dfrac{z^2}{\sigma_{2n+1}^2}\right)}{\left(1 - \dfrac{z^2}{n^2 D^2}\right)^2} = \frac{n^2 L^4}{\sigma_{2n}^2 \, \sigma_{2n+1}^2} \, \frac{(\sigma_{2n}^2 - z^2)\, \sigma_{2n+1}^2 - z^2)}{(n^2 D^2 - z^2)^2} .$$

Now,

(28.13) $$\begin{aligned} \sigma_{2n}^2 \, \sigma_{2n+1}^2 &= \mu_n^2 \, (2n^2 - \mu_n)^2 = (nD + a_n)^2 \, (nD - a_n)^2 \\ &= (n^2 D^2 - a_n^2)^2 \end{aligned}$$

where $|a_n| \leqq D/2$. Hence

(28.14)
$$\prod_1^\infty \frac{n^4 D^4}{\sigma_{2n}^2 \sigma_{2n+1}^2}$$

converges to a finite value. The remaining factor of

(28.15)
$$\prod_1^\infty \frac{\left(1 - \frac{z^2}{\sigma_{2n}^2}\right)\left(1 - \frac{z^2}{\sigma_{2n+1}^2}\right)}{\left(1 - \frac{z^2}{n^2 D^2}\right)^2}$$

is of the form

(28.16)
$$\prod_1^\infty \frac{[(nD - a_n)^2 - z^2]\,[(nD + a_n)^2 - z^2]}{(n^2 L^2 - z^2)^2}$$

where $|a_n| \leqq D/2$. This may be written

(28.17)
$$\prod_1^\infty \left(1 - 2a_n^2 \frac{z^2 + n^2D^2 - a_n^2/2}{(nD + z)^2\,(nD - z)^2}\right).$$

We have $\Im(z) = \delta > 0$. With no restriction, we may consider $\Re(z) = x > 0$. It is easy to see that the uniform boundedness of

(28.18)
$$\int_1^\infty \frac{dt}{(x - Dt)^2 + \delta^2},$$

the fact that only a finite number (independent of z) of the terms

(28.19)
$$\left| 2a_n^2 \frac{z^2 + n^2D^2 - a_n^2/2}{(nD + z)^2\,(nD - z)^2} \right| \geqq \tfrac{1}{2},$$

and the inequality

(28.20)
$$1 - a > \frac{1}{1 + 2a}, \qquad \tfrac{1}{2} > a > 0,$$

prove that

(28.21)
$$c_1 \geqq \left| \prod_1^\infty \left(1 - 2a_n^2 \frac{z^2 + n^2D^2 - a_n^2/2}{(nD + z)^2\,(nD - z)^2}\right) \right| \geqq c_2 .$$

Thus

(28.22)
$$\log \frac{|f(z)|}{|z^2| \prod_1^\infty \left|1 - \frac{z^2}{n^2 D^2}\right|^2} < c \qquad (\Im(z) = \epsilon),$$

and since

$$z \prod_{1}^{\infty} \left| 1 - \frac{z^2}{n^2 L^2} \right| = \frac{1}{\pi} \left| \sin \frac{\pi z}{L} \right|, \tag{28.23}$$

it follows at once that $\log | f(z) |$ is bounded. Hence, by Theorem XXXI, the set of functions $\{1, e^{\pm i\sigma_n x}\}$ is closed over $(-2\pi/L, 2\pi/L)$, but ceases to be closed if we remove any term. Hence if

$$g(x) = \underset{n\to\infty}{\text{l.i.m.}} \sum_{-n}^{n} a_k e^{\pm i\lambda_k x} + a_0 \tag{28.24}$$

vanishes over any interval of length $4\pi/L$, and if, with the exception of a finite number of λ_n's,

$$| \lambda_m - \lambda_n | > L, \tag{28.25}$$

all the coefficients of $g(x)$ will vanish, for otherwise we shall be able to represent some $e^{\pm i\lambda_k x}$ as a limit in the mean in terms of the rest. Since we can replace a finite number of terms of $\{e^{\pm i\sigma_n x}\}$ by the terms $\{e^{\pm i\lambda_n x}\}$ which do not obey (28.25), without affecting the closure properties of the set, we see that the set $\{1, e^{\pm i\lambda_n x}\}$ has been made a subset of a closed and linearly independent sequence. Thus if (28.25) holds for every positive L, when a finite set of λ_n's are dropped, then if (28.24) vanishes over any interval, it vanishes identically. This is a particular case of a gap theorem which we shall take up more thoroughly in the next chapter.

CHAPTER VII

NON-HARMONIC FOURIER SERIES AND A GAP THEOREM

29. A theorem concerning closure. We shall devote the present section to the proof of

THEOREM XXXVII. *Let the set of functions $\{f_n(x)\}$ be normal, orthogonal, and closed over the interval (a, b). Let the functions $g_n(x)$ belong to L_2 over (a, b), and let*

$$\int_a^b \left| \sum_1^N a_n \left(f_n(x) - g_n(x)\right) \right|^2 dx \leqq \theta^2 \sum_1^N |a_n|^2, \tag{29.01}$$

where $\theta < 1$ and where θ is independent of N and a_n. The $\{a_n\}$ are any set of numbers. Then if $f(x)$ belongs to L_2, there exist coefficients b_n such that

$$f(x) = \underset{N\to\infty}{\text{l.i.m.}} \sum_1^N b_n\, g_n(x), \tag{29.02}$$

or in other words, the set $\{g_n(x)\}$ is closed. Moreover, this set is linearly independent.

We shall have

$$\int_a^b |f(x)|^2 dx \leqq (1 + \theta)^2 \sum_1^\infty |b_n|^2, \tag{29.03}$$

and

$$(1 - \theta)^2 \sum_1^\infty |b_n|^2 \leqq \int_a^b |f(x)|^2 dx. \tag{29.04}$$

There exists a set $h_n(x)$ of functions belonging to L_2 such that

$$\int_a^b g_m(x)\, h_n(x)\, dx = \begin{cases} 0 \text{ if } m \neq n; \\ 1 \text{ if } m = n. \end{cases} \tag{29.05}$$

If $f(x)$ belongs to L_2, there exist coefficients c_n such that

$$f(x) = \underset{N\to\infty}{\text{l.i.m.}} \sum_1^N c_n\, h_n(x), \tag{29.06}$$

and

$$\frac{1}{(1+\theta)^2} \sum_1^\infty |c_n|^2 \leqq \int_a^b |f(x)|^2 dx \leqq \frac{1}{(1-\theta)^2} \sum_1^\infty |c_n|^2. \tag{29.07}$$

To begin with,

$$
\begin{aligned}
&\left\{\int_a^b \left|\sum_1^N a_n\, g_n(x)\right|^2 dx\right\}^{1/2} \\
&\leqq \left\{\int_a^b \left|\sum_1^N a_n\, f_n(x)\right|^2 dx\right\}^{1/2} + \left\{\int_a^b \left|\sum_1^N (g_n(x) - f_n(x))\right|^2 dx\right\}^{1/2} \\
(29.08) \qquad &\leqq \left\{\sum_1^N |\, a_n\,|^2\right\}^{1/2} + \left\{\theta^2 \sum_1^N |\, a_n\,|^2\right\}^{1/2} \\
&= (1+\theta)\left\{\sum_1^N |\, a_n\,|^2\right\}^{1/2}.
\end{aligned}
$$

It follows readily that if we consider a particular set $\{a_n\}$ for which

$$
(29.09) \qquad \sum_1^\infty |\, a_n\,|^2 < \infty\,,
$$

then

$$
(29.10) \qquad g(x) = \underset{N\to\infty}{\text{l.i.m.}} \sum_1^N a_n\, g_n(x)
$$

exists, and

$$
(29.11) \qquad \int_a^b |\, g(x)\,|^2\, dx \leqq (1+\theta)^2 \sum_1^\infty |\, a_n\,|^2\,;
$$

for we have only to make use of the fact that by (29.08)

$$
(29.12) \qquad \underset{M,\,N\to\infty}{\text{l.i.m.}} \sum_M^N a_n\, g_n(x) = 0\,.
$$

Now let $f(x)$ be an arbitrary function of L_2, and let

$$
(29.13) \qquad f(x) \sim \sum_1^\infty a_k\, f_k(x)\,.
$$

Then

$$
(29.14) \qquad f^{(1)}(x) \sim \sum_1^\infty a_k(f_k(x) - g_k(x))
$$

will exist and belong to L_2 by (29.01). In general, let us define $f^{(n+1)}(x)$ by mathematical induction. Let us put

$$
(29.15) \qquad f^{(n)}(x) \sim \sum_1^\infty a_k^{(n)}\, f_k(x)\,,
$$

and

$$f^{(n+1)}(x) \sim \sum_1^\infty a_k^{(n)} \left(f_k(x) - g_k(x)\right). \tag{29.16}$$

We shall have

$$f^{(n)}(x) - f^{(n+1)}(x) \sim \sum_1^\infty a_k^{(n)} g_k(x), \tag{29.17}$$

and on summation

$$f(x) - f^{(n+1)}(x) \sim \sum_1^\infty (a_k + a_k^{(1)} + \cdots + a_k^{(n)})\, g_k(x). \tag{29.18}$$

By (29.01) and (29.16),

$$\begin{aligned} \int_a^b |f^{(n+1)}(x)|^2\,dx &\leqq \theta^2 \sum_1^\infty |a_k^{(n)}|^2 = \theta^2 \int_a^b |f^{(n)}(x)|^2\,dx \\ &\leqq \theta^{2(n+1)} \int_a^b |f(x)|^2\,dx. \end{aligned} \tag{29.19}$$

Thus

$$\sum_1^\infty |a_k^{(n)}|^2 \leqq \theta^{2n} \sum_1^\infty |a_k|^2. \tag{29.20}$$

By (29.18) and (29.19),

$$f(x) = \underset{n\to\infty}{\text{l.i.m.}}\ \underset{N\to\infty}{\text{l.i.m.}} \sum_{k=1}^N (a_k + a_k^{(1)} + \cdots + a_k^{(n)})\, g_k(x). \tag{29.21}$$

By the Minkowski inequality and (20.20),

$$\begin{aligned} \left\{\sum_{k=1}^\infty |a_k + a_k^{(1)} + \cdots + a_k^{(n)}|^2\right\}^{1/2} &\leqq \left\{\sum_1^\infty |a_k|^2\right\}^{1/2} \\ &\quad + \left\{\sum_1^\infty |a_k^{(1)}|^2\right\}^{1/2} + \cdots + \left\{\sum_1^\infty |a_k^{(n)}|^2\right\}^{1/2} \\ \leqq (1 + \theta + \cdots + \theta^n)\left\{\sum_1^\infty |a_k|^2\right\}^{1/2} &\leqq \frac{1}{1-\theta}\left\{\sum_1^\infty |a_k|^2\right\}^{1/2}. \end{aligned} \tag{29.22}$$

Thus if we put

$$a_k + a_k^{(1)} + a_k^{(2)} + \cdots = b_k, \tag{29.23}$$

we have

$$\sum_{1}^{\infty} |b_k|^2 \leqq \frac{1}{(1-\theta)^2}\sum_{1}^{\infty} |a_k|^2 = \frac{1}{(1-\theta)^2}\int_a^b |f(x)|^2\,dx\,. \tag{29.24}$$

Let us put

$$g(x) \sim \sum_{1}^{\infty} b_k\, g_k(x)\,. \tag{29.25}$$

This gives us

$$g(x) - f(x) = \underset{n\to\infty}{\text{l.i.m.}}\ \underset{N\to\infty}{\text{l.i.m.}} \sum_{k=1}^{N} (a_k^{(n+1)} + a_k^{(n+2)} + \cdots)\, g_k(x)\,. \tag{29.26}$$

As in (29.22),

$$\left\{\sum_{k=1}^{\infty} |a_k^{(n+1)} + a_k^{(n+2)} + \cdots|^2\right\}^{1/2} \leqq \frac{\theta^{n+1}}{1-\theta}\left\{\sum_{1}^{\infty} |a_k|^2\right\}^{1/2}. \tag{29.27}$$

Hence

$$\int_a^b |g(x) - f(x)|^2\,dx = 0\,, \tag{29.28}$$

and

$$f(x) \sim \sum_{1}^{\infty} b_k\, g_k(x)\,. \tag{29.29}$$

The representation of a function $f(x)$ belonging to L_2 in terms of the functions $g_k(x)$ is unique, provided that the sum of the squares of the moduli of the coefficients converges. Otherwise there will exist a set of b_k's, not all zero, such that

$$\sum b_k\, g_k(x) \sim 0 \tag{29.30}$$

and

$$\infty > \sum |b_k|^2 > 0\,. \tag{29.31}$$

These will give us

$$\begin{aligned} \left\{\int_a^b \left|\sum_{1}^{\infty} b_k f_k(x)\right|^2 dx\right\}^{1/2} &\leqq \left\{\int_a^b \left|\sum_{1}^{\infty} b_k(f_k(x) - g_k(x))\right|^2 dx\right\}^{1/2} \\ &\qquad + \left\{\int_a^b \left|\sum_{1}^{\infty} b_k\, g_k(x)\right|^2 dx\right\}^{1/2} \\ &\leqq \overline{\lim_{N\to\infty}} \left\{\int_a^b \left|\sum_{1}^{N} b_k(f_k(x) - g_k(x))\right|^2 dx\right\}^{1/2} \\ &\leqq \theta\, \overline{\lim_{N\to\infty}} \left\{\sum_{1}^{N} |b_k|^2\right\}^{1/2}. \end{aligned} \tag{29.32}$$

On the other hand,

$$\left\{\int_a^b \left|\sum_1^\infty b_k f_k(x)\right|^2 dx\right\}^{1/2} = \left\{\sum_1^\infty |b_k|^2\right\}^{1/2}. \tag{29.33}$$

This gives us a contradiction.

Let

$$H_n^m(x) = g_n(x) + \sum_1^{n-1} a_k g_k(x) + \sum_{n+1}^m a_k g_k(x), \tag{29.34}$$

and let us choose the a_k's in such a manner that

$$\int_a^b |H_n^m(x)|^2 dx = \text{minimum}. \tag{29.35}$$

The classical theory of orthogonal developments (see §11) will show that

$$H_n(x) = \operatorname*{l.i.m.}_{m\to\infty} H_n^m(x) \tag{29.36}$$

exists and belongs to L_2, and is orthogonal to every $g_k(x)$ other than $g_n(x)$. It is furthermore orthogonal to $g_n(x) - H_n(x)$, which is itself developable in the other g_k's. By (29.04) we have

$$\int_a^b |H_n(x)|^2 dx = \lim_{m\to\infty} \int_a^b |H_n^m(x)|^2 dx \geqq (1-\theta)^2, \tag{29.37}$$

so that no $H_n(x)$ is equivalent to zero. Let us put

$$h_n(x) = \frac{\overline{H_n(x)}}{\int_a^b |H_n(\xi)|^2 d\xi}. \tag{29.38}$$

As we have already seen,

$$\int_a^b g_m(x)\, h_n(x)\, dx = 0 \quad \text{if } m \neq n, \tag{29.39}$$

and on the other hand

$$\begin{aligned}\int_a^b g_n(x)\, h_n(x)\, dx &= \int_a^b g_n(x) \frac{\overline{H_n(x)}}{\int_a^b |H_n(\xi)|^2 d\xi} dx \\ &= \int_a^b (H_n(x) + (g_n(x) - H_n(x))) \frac{\overline{H_n(x)}}{\int_a^b |H_n(\xi)|^2 d\xi} dx.\end{aligned} \tag{29.40}$$

Thus

$$\int_a^b g_n(x)\, h_n(x)\, dx = 1. \tag{29.41}$$

Let

$$\phi_N(x) = \sum_1^N a_n h_n(x), \qquad \psi_N(x) = \sum_1^N \overline{a_n} g_n(x). \tag{29.42}$$

Clearly

$$\int_a^b \phi_N(x)\,\psi_N(x)\,dx = \sum_1^N |a_n|^2, \tag{29.43}$$

and by the Schwarz inequality,

$$\int_a^b |\phi_N(x)|^2\,dx \int_a^b |\psi_n(x)|^2\,dx \geqq \left\{\sum_1^N |a_n|^2\right\}^2. \tag{29.44}$$

Thus by (29.03),

$$\int_a^b |\phi_N(x)|^2\,dx \geqq \frac{\{\sum_1^N |a_n|^2\}^2}{(1+\theta)^2 \sum_1^N |a_n|^2} = \frac{\sum_1^N |a_n|^2}{(1+\theta)^2}. \tag{29.45}$$

Again, let

$$\overline{\phi_N(x)} \sim \sum_1^\infty b_n g_n(x). \tag{29.46}$$

By (29.42), and the Schwarz inequality,

$$\begin{aligned} \int_a^b |\phi_N(x)|^2\,dx &= \sum_1^N a_n b_n \\ &\leqq \left\{\sum_1^N |a_n|^2 \sum_1^N |b_n|^2\right\}^{1/2}. \end{aligned} \tag{29.47}$$

That is, by (29.04),

$$\begin{aligned} \sum_1^N |a_n|^2 &\geqq \frac{\left[\int_a^b |\phi_N(x)|^2\,dx\right]^2}{\sum_1^N |b_n|^2} \\ &\geqq \frac{(1-\theta)^2 \left[\int_a^b |\phi_N(x)|^2\,dx\right]^2}{\int_a^b |\phi_N(x)|^2\,dx} \\ &= (1-\theta)^2 \int_a^b |\phi_N(x)|^2\,dx. \end{aligned} \tag{29.48}$$

The functions $h_n(x)$ are closed. Otherwise there exists a non-null function $f(x)$, belonging to L_2, and such that

$$\int_a^b f(x)\,h_n(x)\,dx = 0 \qquad (n = 1, 2, \cdots). \tag{29.49}$$

On the other hand, we wish to prove that if (29.49) holds valid,

$$f(x) \equiv 0. \tag{29.50}$$

Let us put

$$f(x) \sim \sum_1^\infty b_n g_n(x), \tag{29.51}$$

and

$$h(x) \sim \sum_1^\infty \overline{b_n} h_n(x). \tag{29.52}$$

Then

$$\sum_1^\infty |b_n|^2 = \int_a^b f(x)\, h(x)\, dx = 0. \tag{29.53}$$

This yields our result, and completes the proof of Theorem XXXVII.

A particularly important case is that in which

$$a = -\pi, \qquad b = \pi, \qquad f_n(x) = \frac{1}{(2\pi)^{1/2}} e^{inx} \qquad (-\infty < n < \infty). \tag{29.54}$$

Let us suppose that

$$g_n(x) = \frac{1 + \phi_n(x)}{(2\pi)^{1/2}} e^{inx}. \tag{29.55}$$

Then

$$\int_a^b (f_n(x) - g_n(x))\, (\overline{f_m(x)} - \overline{g_m(x)})\, dx = \frac{1}{2\pi} \int_{-\pi}^{\pi} \phi_n(x)\, \overline{\phi_m(x)}\, e^{i(n-m)x}\, dx. \tag{29.56}$$

If we assume that all the $\phi_n(x)$'s vanish with their first derivatives at $\pm\pi$, and are the twice repeated integrals of their second derivatives, which belong to L_2, we get for (29.56)

$$\begin{aligned} &\left| \frac{1}{2\pi} \int_{-\pi}^{\pi} \frac{e^{i(n-m)x}}{m-n} (\phi_n(x)\overline{\phi_m'(x)} + \phi_n'(x)\overline{\phi_m(x)})\, dx \right| \\ &= \left| \frac{1}{2\pi} \int_{-\pi}^{\pi} \frac{e^{i(n-m)x}}{(n-m)^2} (\phi_n(x)\overline{\phi_m''(x)} + 2\phi_n'(x)\overline{\phi_m'(x)} + \phi_n''(x)\overline{\phi_m(x)})\, dx \right| \\ &\leqq \frac{1}{(n-m)^2} \frac{1}{2\pi} \left\{ \left[\int_{-\pi}^{\pi} |\phi_n(x)|^2\, dx \int_{-\pi}^{\pi} |\phi_m''(x)|^2\, dx \right]^{1/2} \right. \\ &\qquad + 2 \left[\int_{-\pi}^{\pi} |\phi_n'(x)|^2\, dx \int_{-\pi}^{\pi} |\phi_m'(x)|^2\, dx \right]^{1/2} \\ &\qquad \left. + \left[\int_{-\pi}^{\pi} |\phi_n''(x)|^2\, dx \int_{-\pi}^{\pi} |\phi_m(x)|^2\, dx \right]^{1/2} \right\} \\ &\leqq \frac{4\pi}{(n-m)^2} \left\{ \int_{-\pi}^{\pi} |\phi_n''(x)|^2\, dx \int_{-\pi}^{\pi} |\phi_m''(x)|^2\, dx \right\}^{1/2}. \end{aligned} \tag{29.57}$$

Here we have made use of the estimates

$$\begin{aligned}\int_{-\pi}^{\pi} |\phi_n'(x)|^2\,dx &\leqq 2\pi \max |\phi_n'(x)|^2 \\ &\leqq 2\pi \max \left| \int_x^{\pm\pi} \phi_n''(x)\,dx \right|^2 \\ &\leqq 2\pi^2 \int_{-\pi}^{\pi} |\phi_n''(x)|^2\,dx ,\end{aligned} \tag{29.58}$$

and

$$\int_{-\pi}^{\pi} |\phi_n(x)|^2\,dx \leqq 2\pi^2 \int_{-\pi}^{\pi} |\phi_n'(x)|^2\,dx \leqq 4\pi^4 \int_{-\pi}^{\pi} |\phi_n''(x)|^2\,dx . \tag{29.59}$$

Thus if

$$\text{l.u.b.} \int_{-\pi}^{\pi} |\phi_n''(x)|^2\,dx = A , \tag{29.60}$$

we have

$$\begin{aligned}&\int_a^b \left| \sum_{-N}^{N} a_n(f_n(x) - g_n(x)) \right|^2 dx \\ &\leqq 2\pi^3 A \sum_{-N}^{N} |a_n^2| + 4\pi A \sum_{m,n=-N}^{N}{}' \frac{a_m \overline{a_n}}{(m-n)^2} \\ &= A \int_{-\pi}^{\pi} \left| \sum_{-N}^{N} a_n e^{inx} \right|^2 \left(\pi^2 + 2 \sum_{-\infty}^{\infty}{}' \frac{e^{inx}}{n^2} \right) dx \\ &= A \int_{-\pi}^{\pi} \left| \sum_{-N}^{N} a_n e^{inx} \right|^2 \left(\frac{5\pi^2}{3} - 2\pi x + x^2 \right) dx \\ &\leqq 2\pi A \left(\frac{2\pi^2}{3} \right) \sum_{-N}^{N} |a_n|^2 .\end{aligned} \tag{29.61}$$

It follows that if all the $\phi_n(x)$'s are twice repeated integrals of their second derivatives, which belong to L_2, if they all vanish with their first derivatives at $\pm\pi$, and if

$$\text{l.u.b.} \int_{-\pi}^{\pi} |\phi_n''(x)|^2\,dx < \frac{3}{4\pi^3} , \tag{29.62}$$

then (29.01) is fulfilled. There is no reason to suppose that these are by any means the least stringent conditions of the type (which was suggested to one of the authors by Professor Birkhoff) which are sufficient. The type itself is interesting, in that in most of the earlier cases where similar methods of variation of a set of orthogonal functions have been applied, the set corresponding to g_n has actually been asymptotic to the set f_n. It is hoped that the methods of this section may find future applications in the study of special expansion problems.

30. **Non-harmonic Fourier series.** We again let a, b, and the functions $f_n(x)$ be defined as in (29.54), and we put

$$g_n(x) = \frac{1}{(2\pi)^{1/2}} e^{i\lambda_n x}, \tag{30.01}$$

where

$$|\lambda_n - n| \leqq L < \frac{1}{\pi^2}. \tag{30.02}$$

We shall have

$$\begin{aligned} \sum_{-N}^{N} a_n(f_n(x) - g_n(x)) &= \sum_{-N}^{N} \frac{a_n}{(2\pi)^{1/2}} (e^{inx} - e^{i\lambda_n x}) \\ &= ix \sum_{-N}^{N} \frac{a_n}{(2\pi)^{1/2}} \int_{\lambda_n}^{n} e^{iux}\, du \\ &= \frac{ix}{(2\pi)^{1/2}} \int_{-\infty}^{\infty} \psi(u)\, e^{iux}\, du , \end{aligned} \tag{30.03}$$

where $\psi(u)$ is defined as follows:

$$\begin{cases} \text{if } n < \lambda_n , & \psi(u) = -a_n \quad \text{over } (n, \lambda_n) ; \\ \text{if } \lambda_n < n , & \psi(u) = a_n \quad \text{over } (\lambda_n, n) ; \\ \text{outside these intervals } \psi(u) = 0 . \end{cases} \tag{30.04}$$

By Plancherel's theorem,

$$\begin{aligned} \int_{-\infty}^{\infty} \frac{1}{x^2} \left| \sum_{-N}^{N} a_n(f_n(x) - g_n(x)) \right|^2 dx &= \int_{-\infty}^{\infty} |\psi(u)|^2\, du \\ &= \sum_{-N}^{N} |n - \lambda_n|\, |a_n|^2 \leqq L \sum_{-N}^{N} |a_n|^2 . \end{aligned} \tag{30.05}$$

The inequality

$$\int_{-\infty}^{\infty} \frac{1}{x^2} \left| \sum_{-N}^{N} a_n(f_n(x) - g_n(x)) \right|^2 dx \geqq \int_{-\pi}^{\pi} \frac{1}{\pi^2} \left| \sum_{-N}^{N} a_n(f_n(x) - g_n(x)) \right|^2 dx \tag{30.06}$$

is obvious. Thus we have

$$\int_{-\pi}^{\pi} \left| \sum_{-N}^{N} a_n(f_n(x) - g_n(x)) \right|^2 dx \leqq \pi^2 L \sum_{-N}^{N} |a_n|^2 , \tag{30.07}$$

which establishes (29.01) in this case. Thus we may apply Theorem XXXVII, and we see that the set of functions $\{e^{i\lambda_n x}\}$ are closed L_2 over $(-\pi, \pi)$. This is a mere confirmation of part of Theorem XXXIV; what is strictly new is that if $f(x)$ belongs to L_2 over $(-\pi, \pi)$ we may write

$$f(x) = \underset{N\to\infty}{\text{l.i.m.}} \sum_{-N}^{N} a_n e^{i\lambda_n x}, \tag{30.08}$$

where

$$\int_{-\pi}^{\pi} |f(x)|^2 dx \leqq \text{const.} \sum_{-\infty}^{\infty} |a_n|^2 \tag{30.09}$$

and

$$\sum_{-\infty}^{\infty} |a_n|^2 \leqq \text{const.} \int_{-\pi}^{\pi} |f(x)|^2 dx. \tag{30.10}$$

Defining $f(x)$ as in (30.08), let us put

$$f_\xi(x) = \underset{N\to\infty}{\text{l.i.m.}} \sum_{-N}^{N} a_n e^{i\lambda_n \xi} e^{i\lambda_n x}. \tag{30.11}$$

It follows from (30.09) and (30.10) that this function will exist and belong to L_2, and by comparing it for different values of ξ, we see that we may write it in the form

$$f_\xi(x) = f_b(x + \xi) \qquad [-\infty < x < \infty]. \tag{30.12}$$

By (30.09) we shall have

$$\int_{-\pi}^{\pi} |f_b(x+\xi)|^2 dx \leqq \text{const.} \sum_{-\infty}^{\infty} |a_n e^{i\lambda_n \xi}|^2 = \text{const.} \sum_{-\infty}^{\infty} |a_n|^2, \tag{30.13}$$

and by (30.10),

$$\int_{-\pi}^{\pi} |f_b(x+\xi)|^2 dx \leqq \text{const.} \int_{-\pi}^{\pi} |f(x)|^2 dx. \tag{30.14}$$

Let $f_1(x)$ be the periodic function of period 2π which coincides with $f(x)$ over $(-\pi, \pi)$, and let

$$g(x) = f_a(x) - f_b(x). \tag{30.15}$$

By Minkowski's theorem, we have

$$\int_{-\pi}^{\pi} |g(x+\xi)|^2 dx \leqq \text{const.}, \tag{30.16}$$

and by the definition of $g(x)$, we have (except at most over a null set)

$$g(x) = 0 \qquad (-\pi < x < \pi). \tag{30.17}$$

Thus

$$\frac{1}{2B}\int_{-B}^{B} |g(x)|^2\,dx \tag{30.18}$$

is bounded. It follows that

$$\begin{aligned}\int_{\pi}^{B}\frac{|g(x)|^2\,dx}{x^2} &= \int_{\pi}^{B}\frac{1}{x^2}\,d\int_{-x}^{x}|g(\xi)|^2\,d\xi \\ &= \frac{1}{B^2}\int_{-B}^{B}|g(x)|^2\,dx + \int_{\pi}^{B}\frac{2dx}{x^3}\int_{-x}^{x}|g(\xi)|^2\,d\xi \\ &\leqq \frac{c}{B} + \int_{\pi}^{B}\frac{c\,dx}{x^2} = c,\end{aligned} \tag{30.19}$$

where c represents various constants. Hence $g(x)/x$ belongs to L_2. By Plancherel's theorem, it has a Fourier transform

$$G(u) = \frac{1}{(2\pi)^{1/2}}\operatorname*{l.i.m.}_{A\to\infty}\int_{-A}^{A}\frac{g(x)\,e^{-iux}\,dx}{x}, \tag{30.20}$$

likewise belonging to L_2, and

$$\frac{g(x)}{x} = \frac{1}{(2\pi)^{1/2}}\operatorname*{l.i.m.}_{A\to\infty}\int_{-A}^{A}G(u)\,e^{iux}\,du. \tag{30.21}$$

It follows that

$$\frac{g(x)\,(1 - e^{iwx})}{x} = \frac{1}{(2\pi)^{1/2}}\operatorname*{l.i.m.}_{A\to\infty}\int_{-A}^{A}[G(u) - G(u - w)]\,e^{iux}\,du. \tag{30.22}$$

On the other hand, over any infinite range,

$$g(x) = f_a(x) - f_b(x) = \operatorname*{l.i.m.}_{N\to\infty}\sum_{-N}^{N}(a_n\,e^{inx} - b_n\,e^{i\lambda_n x}), \tag{30.23}$$

so that over $(-\infty, \infty)$,

$$\frac{g(x)\,(1 - e^{iwx})}{x} = \operatorname*{l.i.m.}_{N\to\infty}\sum_{-N}^{N}\left(ia_n\int_{n+w}^{n}e^{iux}\,du - ib_n\int_{\lambda_n+w}^{\lambda_n}e^{iux}\,du\right). \tag{30.24}$$

It will be seen at once that since $\sum_{-\infty}^{\infty}|a_n|^2$ and $\sum_{-\infty}^{\infty}|b_n|^2$ converge, (30.24) represents $g(x)\,(1 - e^{iwx})/x$ as the Fourier transform of a function of L_2, which must be almost everywhere the same as the transform given by (30.22).

This is only possible if

$$G(u) + \sum_{1<n<u} ia_n\,(2\pi)^{1/2} - \sum_{1<\lambda_n<u} ib_n\,(2\pi)^{1/2} \sim K = \text{const.}\quad (u > 0). \tag{30.25}$$

Let us notice that this and a similar result for negative u lead us to

$$G(u) = \text{const.}\qquad [\,|n + \tfrac{1}{2} - u| < \tfrac{1}{10}] \tag{30.26}$$

except for a null set in each such interval, which we may take to be empty. Since $G(u)$ belongs to L_2, this gives us

$$\lim_{n\to\pm\infty} G(n+\tfrac{1}{2}) = 0 \tag{30.27}$$

and in view of the fact that a_n and b_n tend to zero as $n \to \pm\infty$, we may so define $G(u)$ that

$$\lim_{u\to\pm\infty} G(u) = 0\,. \tag{30.28}$$

Thus by (30.25) and the analogous result for negative u,

$$\sum_{2}^{\infty} (ia_n(2\pi)^{1/2} - ib_n(2\pi)^{1/2}) = K = -\sum_{-\infty}^{1} (ia_n(2\pi)^{1/2} - ib_n(2\pi)^{1/2})\,, \tag{30.29}$$

which permits us to write

$$G(u) = \sum_{u<n} ia_n(2\pi)^{1/2} - \sum_{u<\lambda_n} ib_n(2\pi)^{1/2}\,. \tag{30.30}$$

In this summation, with the possible exception of the last term, the terms $ia_m(2\pi)^{1/2} - ib_m(2\pi)^{1/2}$ are bracketed.

Let x lie within the interval $(-\pi+\epsilon, \pi-\epsilon)$. Let us put

$$F_w(x) = \frac{\sin ux}{x} \quad (|x| \geqq \epsilon/2)\,; \quad F_w(x) = 0 \quad (|x| < \epsilon/2)\,; \tag{30.31}$$

$$H_w(u) = \underset{A\to\infty}{\text{l.i.m.}} \frac{1}{(2\pi)^{1/2}} \int_{-A}^{A} F_w(x)\, e^{iux}\, dx\,. \tag{30.32}$$

We have

$$H_w(u) = \frac{1}{2i} (\phi(u+w) + \phi(u-w))\,, \tag{30.33}$$

where

$$\phi(u) = \underset{A\to\infty}{\text{l.i.m.}} \frac{1}{(2\pi)^{1/2}} \left[\int_{-A}^{-\epsilon/2} + \int_{\epsilon/2}^{A}\right] \frac{e^{iux}\, dx}{x}\,. \tag{30.34}$$

From this it readily results that

$$\begin{aligned} \overline{\lim_{w\to\infty}} \int_{-A}^{A} |H_w(u)|^2\, du &\leqq \overline{\lim_{w\to\infty}} \int_{-A+w}^{A+w} |\phi(u)|^2\, du \\ &\quad + \overline{\lim_{w\to\infty}} \int_{-A-w}^{A-w} |\phi(u)|^2\, du = 0\,. \end{aligned} \tag{30.35}$$

On the other hand,

$$\int_{-\infty}^{\infty} |H_w(u)|^2\, du \leqq 2 \int_{\epsilon/2}^{\infty} \frac{dx}{x^2} = \frac{4}{\epsilon}\,. \tag{30.36}$$

Over the interval $(-\pi + \epsilon, \pi - \epsilon)$, we have

$$\left| \frac{1}{\pi} \int_{-\infty}^{\infty} \frac{g(\xi)}{\xi} \frac{\sin w\,(x-\xi)}{x-\xi}\, d\xi \right| = \left| \frac{1}{\pi} \int_{-\infty}^{\infty} \frac{g(\xi)}{\xi} F_w(x-\xi)\, d\xi \right|$$

$$= \left| \frac{1}{\pi} \int_{-\infty}^{\infty} G(-u)\, H_w(u)\, e^{-iux}\, du \right|$$

$$\leqq \frac{1}{\pi} \left\{ \int_{-A}^{A} |\, G(u)\,|^2\, du \int_{-A}^{A} |\, H_w(u)\,|^2\, du \right\}^{1/2}$$

$$(30.37) \qquad + \frac{1}{\pi} \left\{ \left[\int_{A}^{\infty} + \int_{-\infty}^{-A} \right] |\, G(u)\,|^2\, du \left[\int_{A}^{\infty} + \int_{-\infty}^{-A} \right] |H_w(u)\,|^2\, du \right\}^{1/2}$$

$$\leqq \text{const.} \left\{ \int_{-A}^{A} |\, H_w(u)\,|^2\, du \right\}^{1/2}$$

$$+ \text{ const.} \left\{ \left[\int_{A}^{\infty} + \int_{-\infty}^{-A} \right] |\, G(u)\,|^2\, du \right\}^{1/2}.$$

In the last line of this formula, we may first take A so large that the second term does not exceed $\eta/2$, and then take w so large that the first term does not exceed $\eta/2$. It follows at once since η is arbitrary that we have

$$(30\ 38) \qquad \lim_{w \to \infty} \frac{1}{\pi} \int_{-\infty}^{\infty} \frac{g(\xi)}{\xi} \frac{\sin w(x-\xi)}{x-\xi}\, d\xi = 0$$

uniformly over $(-\pi + \epsilon, \pi - \epsilon)$.

Now since

$$(30.39) \qquad \frac{1}{2} \int_{-w}^{w} e^{iux}\, du = \frac{\sin wx}{x}; \qquad \frac{1}{\pi} \int_{-\infty}^{\infty} \frac{\sin wx}{x} e^{-iux}\, dx = \begin{cases} 1 \text{ if } |\, u\,| < w; \\ 0 \text{ if } |\, u\,| > w; \end{cases}$$

it follows by the Parseval theorem that

$$\frac{1}{\pi} \int_{-\infty}^{\infty} \frac{g(\xi)}{\xi} \frac{\sin w(x-\xi)}{x-\xi}\, d\xi = \frac{1}{(2\pi)^{1/2}} \int_{-w}^{w} G(u)\, e^{iux}\, du$$

$$(30.40) \qquad = \frac{1}{(2\pi)^{1/2}} G(w) \frac{e^{iwx}}{ix} - \frac{1}{(2\pi)^{1/2}} G(-w) \frac{e^{-iwx}}{ix}$$

$$- \frac{1}{(2\pi)^{1/2}} \int_{-w}^{w} \frac{e^{iux}\, dG(u)}{ix}.$$

However,

$$(30.41) \qquad - \frac{1}{(2\pi)^{1/2}} \int_{-w}^{w} \frac{e^{iux}\, dG(u)}{ix} = \frac{1}{x} \left[\sum_{-[w]}^{[w]} a_n\, e^{inx} - \sum_{w > \lambda_n > -w} b_n\, e^{i\lambda_n x} \right].$$

Since $G(w)$ tends to zero as $w \to \pm\infty$ and $b_n \to 0$ as $n \to \pm\infty$,

$$\frac{x}{\pi}\int_{-\infty}^{\infty}\frac{g(\xi)}{\xi}\frac{\sin w(x-\xi)}{x-\xi}\,d\xi = \sum_{-[w]}^{[w]}(a_n e^{inx} - b_n e^{i\lambda_n x}) + o(1), \tag{30.42}$$

and by (30.38) we have uniformly

$$\lim_{N\to\infty}\sum_{-N}^{N}(a_n e^{inx} - b_n e^{i\lambda_n x}) = 0 \tag{30.43}$$

over $(-\pi+\epsilon, \pi-\epsilon)$. We thus obtain

THEOREM XXXVIII. *Let*

$$|\lambda_n - n| \leqq L < \frac{1}{\pi^2} \qquad [n = 0, 1, 2, \cdots; -1, -2, \cdots]. \tag{30.44}$$

Then the set of functions $\{e^{i\lambda_n x}\}$ *is closed* L_2 *over* $(-\pi, \pi)$, *and admits a unique closed normal biorthogonal set* $\{h_n(x)\}$. *If* $f(x)$ *is any function belonging to* L_2 *over* $(-\pi, \pi)$, *the series*

$$\sum_{-\infty}^{\infty}\left\{\frac{e^{inx}}{2\pi}\int_{-\pi}^{\pi}f(\xi)\,e^{-in\xi}\,d\xi - e^{i\lambda_n x}\int_{-\pi}^{\pi}f(\xi)\,h_n(\xi)\,d\xi\right\} \tag{30.45}$$

converges uniformly to zero over any interval $(-\pi+\epsilon \leqq x \leqq \pi-\epsilon)$, *and over any such interval the convergence and summability properties of the series*

$$\sum_{-\infty}^{\infty} e^{i\lambda_n x}\int_{-\pi}^{\pi}f(\xi)\,h_n(\xi)\,d\xi \tag{30.46}$$

are the same as those of the ordinary Fourier series

$$\sum_{-\infty}^{\infty}\frac{e^{inx}}{2\pi}\int_{-\pi}^{\pi}f(\xi)\,e^{-in\xi}\,d\xi. \tag{30.47}$$

We still do not know whether this equivalence holds for functions of L_1 in general, or indeed, what any of the properties of such series are. We definitely do know that the properties of equivalence which we have established do not hold for absolute convergence, for the alteration of a single one of the functions e^{inx} is in general enough to upset these properties.

We have no proof that the constant $1/\pi^2$ which occurs in (30.44) is a best possible constant. In (30.06), in replacing

$$\int_{-\infty}^{\infty}\frac{1}{x^2}\left|\sum_{-N}^{N}a_n(f_n(x) - g_n(x))\right|^2 dx \tag{30.48}$$

by

$$\frac{1}{\pi^2}\int_{-\pi}^{\pi}\left|\sum_{-N}^{N} a_n(f_n(x) - g_n(x))\right|^2 dx \tag{30.49}$$

we have gone through the intermediate stage

$$\int_{-\pi}^{\pi}\frac{1}{x^2}\left|\sum_{-N}^{N} a_n(f_n(x) - g_n(x))\right|^2 dx. \tag{30.50}$$

It is perfectly possible to produce examples to show that the transition from (30.50) to (30.49) is in some sense a best possible one, in that all the large values of $\sum_{-N}^{N} a_n(f_n(x) - g_n(x))$ may be concentrated near $\pm\pi$. The transition from (30.48) to (30.50) may however involve a loss of accuracy harder to take account of.*

Theorem XXXVIII proves both more and less than Theorem XXXIII, which we have established by quite different methods. The interval of Theorem XXXVIII is narrower than that of Theorem XXXIII, but on the other hand, we have given up the conditions that $\lambda_0 = 0$ and that $\lambda_{-n} = -\lambda_n$. We have also proved far more than simple closure.

The biorthogonal set of functions $\{h_n(x)\}$ deserves a certain amount of attention. We shall confine ourselves to the case in which $\lambda_0 = 0$, $\lambda_{-n} = -\lambda_n$, or to the closely related case differing from this in only a finite number of λ_k's. This second case is so like the first that we do not need to take up its properties separately.

Let us then put

$$G(z) = z\prod_{1}^{\infty}\left(1 - \frac{z^2}{\lambda_n^2}\right). \tag{30.51}$$

By Theorem XXX, $G(z)$ will not belong to L_2 along the real axis, but $G(z)/z$ will so belong. Thus the functions

$$h_n(x) = \frac{1}{2\pi G'(\lambda_n)}\ \underset{A\to\infty}{\text{l.i.m.}}\int_{-A}^{A}\frac{G(u)}{u - \lambda_n}e^{-iux}\,du \tag{30.52}$$

will all belong to L_2. By (27.20) we shall have, except over a null set,

$$h_n(x) = 0 \qquad [|x| > \pi], \tag{30.53}$$

and since the functions $G(u)/(G'(\lambda_n)(u - \lambda_n))$ are analytic, taking on the value 1 at λ_n, and the value 0 at $\lambda_m (m \neq n)$,

$$\int_{-\pi}^{\pi} h_m(x)\,e^{i\lambda_n x}\,dx = \int_{-\infty}^{\infty} h_m(x)\,e^{i\lambda_n x}\,dx = \begin{cases}1 \text{ if } m = n;\\ 0 \text{ if } m \neq n.\end{cases} \tag{30.54}$$

* The constant $1/\pi^2$ has been recently improved by Dr. H. Malin in his dissertation at the Massachusetts Institute of Technology. The best constant is not yet known.

Thus the functions $h_n(x)$ are the functions represented by the same symbol in Theorem XXXVIII. They are a closed set over $(-\pi, \pi)$, and if $f(x)$ belongs to L_2,

$$f(x) \sim \sum_{-\infty}^{\infty} h_n(x) \int_{-\pi}^{\pi} f(\xi)\, e^{i\lambda_n \xi}\, d\xi . \tag{30.55}$$

In this case,

$$\frac{1}{C} \sum_{-\infty}^{\infty} \left| \int_{-\pi}^{\pi} f(\xi)\, e^{i\lambda_n\xi}\, d\xi \right|^2 \leqq \int_{-\pi}^{\pi} | f(x) |^2\, dx \leqq C \sum_{-\infty}^{\infty} \left| \int_{-\pi}^{\pi} f(\xi)\, e^{i\lambda_n\xi}\, d\xi \right|^2 , \tag{30.56}$$

as Theorem XXXVII shows.

By a Fourier transformation, we have

$$\frac{1}{C_1} \sum_{-\infty}^{\infty} | a_n |^2 \leqq \int_{-\infty}^{\infty} \left| \underset{N\to\infty}{\text{l.i.m.}} \sum_{-N}^{N} \frac{a_n\, G(u)}{G'(\lambda_n)(u - \lambda_n)} \right|^2 du \leqq C_1 \sum_{-\infty}^{\infty} | a_n |^2 \tag{30.57}$$

where

$$a_n = \int_{-\pi}^{\pi} f(\xi)\, e^{i\lambda_n\xi}\, d\xi , \tag{30.58}$$

in the extended sense that if either side of the inequalities exists, the other will exist and satisfy the inequalities.

The expression

$$\underset{N\to\infty}{\text{l.i.m.}} \sum_{-N}^{N} \frac{a_n\, G(u)}{G'(\lambda_n)(u - \lambda_n)} \tag{30.59}$$

may be regarded as the Lagrange interpolation formula for a function assuming the values a_n for the arguments λ_n. We may summarize our results, as far as they concern interpolation, as follows:

THEOREM XXXIX. *Let*

$$| \lambda_n - n | \leqq L < \frac{1}{\pi^2} \qquad (n = 1, 2, \cdots); \qquad \lambda_0 = 0; \qquad \lambda_{-n} = -\lambda_n . \tag{30.60}$$

Let (30.59) *be called the Lagrange interpolation formula for a function assuming the values a_n at the points λ_n. Then the class of all functions defined by such Lagrange interpolation formulas for which*

$$\sum_{-\infty}^{\infty} | a_n |^2 < \infty \tag{30.61}$$

is the same as the class of all entire functions $\phi(u)$ belonging to L_2 along the real axis, and satisfying

$$\overline{\lim_{r\to\infty}} \frac{1}{\pi r} \log | \phi(re^{i\theta}) | \leqq 1 ; \tag{30.62}$$

or along the real axis, is the same as the class of all functions of L_2 with Fourier transforms vanishing outside $(-\pi, \pi)$. *Furthermore* (30.57) *and* (30.58) *hold good.*

In establishing this theorem, we have used Theorem V.

31. **A new class of almost periodic functions.** We shall say that $f(x)$, a (generally complex-valued) function of the real argument $x(-\infty < x < \infty)$ is *pseudoperiodic* if $f(x)$ belongs to L_2 over every finite range, and there exist two positive numbers, A and B, such that if $(a_1, \cdots, a_n)$ is any set of complex numbers and $(b_1, \cdots, b_n)$ any set of real numbers, and x and y are real,

$$\frac{\int_x^{x+A} \left| \sum_1^n a_k f(\xi + b_k) \right|^2 d\xi}{\int_y^{y+A} \left| \sum_1^n a_k f(\xi + b_k) \right|^2 d\xi} < B. \tag{31.01}$$

We shall call the greatest lower bound of A for all $B > 0$ the *pseudoperiod* of $f(x)$.

THEOREM XL. *The class of pseudoperiodic functions is equivalent to the class of all functions not equivalent to zero which belong to the class Stepanoff* 2,* *whose characteristic frequencies* $\{\lambda_m\}$ *are so spaced that*

$$\text{g.l.b.} \mid \lambda_m - \lambda_n \mid > 0. \tag{31.02}$$

Here a function $f(x)$ of L_2 is said to belong to class Stepanoff 2 with characteristic frequencies $\{\lambda_n\}$, if given any ϵ, we can find an integer N and a polynomial

$$P_\epsilon(x) = \sum_1^N a_n e^{i\lambda_n x} \tag{31.03}$$

such that for every x

$$\int_x^{x+1} \mid P_\epsilon(\xi) - f(\xi) \mid^2 d\xi < \epsilon. \tag{31.04}$$

Let $f(x)$ be such a function, and let (31.02) be positive and greater than L. Then the functions

$$\left\{ e^{i\lambda_n x} \frac{\sin \frac{Lx}{2}}{x} \right\}$$

have Fourier transforms

* W. Stepanoff, *Sur quelques généralisations des fonctions presque périodiques*, Comptes Rendus, vol. 181, pp. 90–92.

$$(31.05)\qquad \frac{1}{(2\pi)^{1/2}}\int_{-\infty}^{\infty} e^{iux}e^{i\lambda_n x}\frac{\sin\frac{Lx}{2}}{x}\,dx = \begin{cases}\left(\frac{\pi}{2}\right)^{1/2} & \text{if } |\lambda_n + u| < \frac{L}{2},\\ 0 & \text{if } |\lambda_n + u| > \frac{L}{2},\end{cases}$$

which differ from zero over non-overlapping ranges, and are orthogonal. Thus using Plancherel's theorem,

$$(31.06)\qquad \int_{-\infty}^{\infty} |P_\epsilon(x)|^2 \frac{\sin^2\frac{Lx}{2}}{x^2}\,dx = \sum_1^N |a_n|^2 \frac{\pi L}{2}.$$

Hence

$$(31.07)\qquad \int_{-\pi/L}^{\pi/L} |P_\epsilon(x)|^2\,dx \leqq \frac{\pi^3}{2L}\sum_1^N |a_n|^2,$$

and in general

$$(31.071)\qquad \int_{-\pi/L+y}^{\pi/L+y} |P_\epsilon(x)|^2\,dx \leqq \frac{\pi^3}{2L}\sum_1^N |a_n|^2.$$

Now let $\sum_1^N |a_n|^2$ be given. By (31.071), if $A > \pi/L$,

$$(31.08)\qquad \int_{-A}^{A} |P_\epsilon(x)|^2\,dx \leqq \pi^2 A \sum_1^N |a_n|^2.$$

Thus

$$(31.09)\qquad \begin{aligned}\left|\left[\int_A^\infty + \int_{-\infty}^{-A}\right]\frac{|P_\epsilon(x)|^2}{x^2}\,dx\right| &= \left|\int_A^\infty \frac{1}{x^2}\,d\int_{-x}^{x} |P_\epsilon(\xi)|^2\,d\xi\right|\\ &= \left|-\frac{1}{A^2}\int_{-A}^{A}|P_\epsilon(\xi)|^2\,d\xi + 2\int_A^\infty \frac{dx}{x^2}\frac{1}{x}\int_{-x}^{x}|P_\epsilon(\xi)|^2\,d\xi\right|\\ &\leqq 2\pi^2\sum_1^N |a_n|^2\int_A^\infty \frac{dx}{x^2} = \frac{2\pi^2}{A}\sum_1^N |a_n|^2.\end{aligned}$$

Hence using (31.06)

$$(31.10)\qquad \int_{-A}^{A} |P_\epsilon(x)|^2\frac{\sin^2\frac{Lx}{2}}{x^2}\,dx \geqq \left[\frac{\pi L}{2} - \frac{2\pi^2}{A}\right]\sum_1^N |a_n|^2,$$

and if $c > 0$,

$$\int_{-4\pi^2/(\pi L-2c)}^{4\pi^2/(\pi L-2c)} |P_\epsilon(x)|^2 dx \geqq \frac{4}{L^2}\int_{-4\pi^2/(\pi L-2c)}^{4\pi^2/(\pi L-2c)} |P_\epsilon(x)|^2 \frac{\sin^2 \frac{Lx}{2}}{x^2} dx$$

(31.11)
$$\geqq \frac{4c}{L^2}\sum_1^n |a_n|^2$$

$$\geqq \frac{8c}{L^2\pi^2}\frac{\pi L - 2c}{4\pi^2}\int_{-4\pi^2/(\pi L-2c)+y}^{4\pi^2/(\pi L-2c)+y} |P_\epsilon(x)|^2 dx .$$

We have thus established (31.01) for

(31.12)
$$A = \frac{8\pi^2}{\pi L - 2c}, \qquad B^{-1} = \frac{2c}{\pi^4 L^2}(\pi L - 2c),$$

and for $P_\epsilon(x)$ and its translations and their linear combinations, instead of $f(x)$ and the functions similarly obtained from it. By the Minkowski inequality the result may be transferred to $f(x)$ itself. By (31.12) the pseudoperiod of $f(x)$ is not greater than $8\pi/L$.

On the other hand, let $f(x)$ be pseudoperiodic, with A and B as in (31.01). Over any finite range, if $K(x)$ belongs to L_2 and vanishes outside $(-D, D)$, $\int_{-\infty}^{\infty} K(x-\xi)f(\xi)\,d\xi$ can be approximated to in the mean by a polynomial $\sum_{-N}^{N} a_n f(x-b_n)$* and hence if $(x, x+A)$ and $(y, y+A)$ be within this range,

(31.13)
$$\frac{\displaystyle\int_x^{x+A}\left|\int_{-\infty}^{\infty} K(\eta-\xi)f(\xi)\,d\xi\right|^2 d\eta}{\displaystyle\int_y^{y+A}\left|\int_{-\infty}^{\infty} K(\eta-\xi)f(\xi)\,d\xi\right|^2 d\eta} \leqq B.$$

Since the range is arbitrary, this holds for all real x and y. In particular, if

(31.14)
$$f_\epsilon(x) = \frac{1}{\epsilon}\int_{x-\epsilon}^{x+\epsilon}\left|1 - \frac{|\xi - x|}{\epsilon}\right| f(\xi)\,d\xi,$$

---

*
$$\int_{-\infty}^{\infty} K(x-\xi)\,f(\xi)\,d\xi = \int_{-D+x}^{D+x} K(x-\xi)\,f(\xi)\,d\xi\,;$$

therefore over any finite range of x we can use $f_N(x) = f(x), |x| \leqq N; f_N(x) = 0, |x| > N$. Let the transform of $f_N(x)$ be $g_N(u)$ and of $K(x)$ be $k(u)$. Then we want to approximate in the mean to

$$\int_{-\infty}^{\infty} k(u)\,g_N(u)e^{iux}\,du = \int_{-D+x}^{D+x} K(x-\xi)\,f(\xi)\,d\xi$$

over any range $|x| < N - D$. If we now use the Plancherel theorem and the Minkowski inequality and we expand $k(u)$ into a Fourier Series over $(-c, c)$ so that

$$k(u) \sim \sum_{-\infty}^{\infty} a_n\,e^{i\pi nu/c}$$

over this range, then we have the L_2 approximation in the desired form $\sum_{-n}^{n} a_n f(x + b_n)$.

then

$$\frac{\int_x^{x+A} |f_\epsilon(\xi)|^2 \, d\xi}{\int_y^{y+A} |f_\epsilon(\xi)|^2 \, d\xi} \leqq B; \tag{31.15}$$

$$\frac{\int_x^{x+A} |f'_\epsilon(\xi)|^2 \, d\xi}{\int_y^{y+A} |f'_\epsilon(\xi)|^2 \, d\xi} \leqq B; \tag{31.16}$$

and it results directly from (31.01) that

$$\frac{\int_x^{x+A} |f''_\epsilon(\xi)|^2 \, d\xi}{\int_y^{y+A} |f''_\epsilon(\xi)|^2 \, d\xi} < B. \tag{31.17}$$

Hence if $w < A$,

$$|f'_\epsilon(x+w) - f'_\epsilon(x)|^2 = \left| \int_x^{x+w} f''_\epsilon(\xi) \, d\xi \right|^2 < A \int_x^{x+w} |f''_\epsilon(\xi)|^2 \, d\xi < C. \tag{31.18}$$

By (31.16),

$$\int_x^{x+A} |f'_\epsilon(\xi)|^2 \, d\xi \leqq C, \tag{31.19}$$

and it results that for some ξ between x and $x + A$, $f'_\epsilon(\xi) \leqq C$. Hence by (31.18)

$$|f'_\epsilon(x)| < C. \tag{31.20}$$

Similarly, by (31.15),

$$|f_\epsilon(x)| < C. \tag{31.21}$$

Thus the functions $f_\epsilon(x + \sigma_n)$ are equally bounded and equicontinuous, and over any finite range, if $\{\sigma_n\}$ is any sequence of real numbers, it contains a subsequence $\{\mu_n\}$ such that

$$\lim_{n \to \infty} f_\epsilon(x + \mu_n) = g_\epsilon(x) \tag{31.22}$$

exists uniformly. By (31.01), over $-\infty < x < \infty$,

$$\lim_{m,n \to \infty} \int_x^{x+A} |f_\epsilon(\xi + \mu_m) - f_\epsilon(\xi + \mu_n)|^2 \, d\xi = 0, \tag{31.23}$$

so that uniformly over any range $(x, x + A)$,

$$g_\epsilon(x) = \underset{n \to \infty}{\text{l.i.m.}} f_\epsilon(x + \mu_n) \tag{31.24}$$

exists. We shall have uniformly

$$(31.25)\qquad \frac{1}{2\epsilon}\int_{x-\epsilon}^{x+\epsilon} g_\epsilon(\xi)\,d\xi = \lim_{n\to\infty}\frac{1}{2\epsilon}\int_{x-\epsilon}^{x+\epsilon} f_\epsilon(\xi+\mu_n)\,d\xi\,.$$

That is,

$$(31.26)\qquad \phi_\epsilon(x) = \frac{1}{2\epsilon}\int_{x-\epsilon}^{x+\epsilon} f_\epsilon(\xi)\,d\xi$$

is normal* in the sense of Bochner, in that every sequence $\phi_\epsilon(x + b_n)$ contains a subsequence with a uniform limit over $(-\infty, \infty)$, and hence is almost periodic. As a consequence†

$$(31.27)\qquad a_\Lambda = \lim_{T\to\infty}\frac{1}{T}\int_x^{x+T}\phi_\epsilon(\xi)\,e^{i\Lambda\xi}\,d\xi$$

will exist uniformly in x for all Λ, and will differ from 0 for only a denumerable set $\{\Lambda_n\}$ of values of Λ. By (31.13),

(31.28)

$$\frac{\displaystyle\int_y^{y+A}\left|\frac{1}{\alpha_{\Lambda_m}T}\int_x^{x+T}\phi_\epsilon(\xi)\,e^{i\Lambda_m(\xi-x)}\,d\xi - \frac{1}{\alpha_{\Lambda_n}T}\int_x^{x+T}\phi_\epsilon(\xi)\,e^{i\Lambda_n(\xi-x)}\,d\xi\right|^2 dx}{\displaystyle\int_z^{z+A}\left|\frac{1}{\alpha_{\Lambda_m}T}\int_x^{x+T}\phi_\epsilon(\xi)\,e^{i\Lambda_m(\xi-x)}\,dx - \frac{1}{\alpha_{\Lambda_n}T}\int_x^{x+T}\phi_\epsilon(\xi)\,e^{i\Lambda_n(\xi-x)}\,d\xi\right|^2 dx} \leqq B,$$

and by (31.27)

$$(31.29)\qquad \frac{\displaystyle\int_y^{y+A}|e^{-i\Lambda_m x} - e^{-i\Lambda_n x}|^2\,dx}{\displaystyle\int_z^{z+A}|e^{-i\Lambda_m x} - e^{-i\Lambda_n x}|^2\,dx} \leqq B.$$

That is,

$$(31.30)\qquad \frac{\displaystyle\int_y^{y+A}(1-\cos(\Lambda_m-\Lambda_n)x)\,dx}{\displaystyle\int_z^{z+A}(1-\cos(\Lambda_m-\Lambda_n)x)\,dx} \leqq B,$$

whence if $\Lambda_m > \Lambda_n$ and $\Lambda_m - \Lambda_n$ is small enough and we let

$$y = \frac{\pi}{\Lambda_m - \Lambda_n} - \frac{A}{2}, \qquad\qquad z = -\frac{A}{2},$$

* A. S. Besicovitch, *Almost Periodic Functions*, Cambridge, 1932, p. 10.

† Ibid., p. 15.

we see that

$$\frac{1 + \cos(\Lambda_m - \Lambda_n)\dfrac{A}{2}}{1 - \cos(\Lambda_m - \Lambda_n)\dfrac{A}{2}} < B, \tag{31.31}$$

from which we get

$$(\Lambda_m - \Lambda_n)\frac{A}{2} > B^{-1/2}. \tag{31.32}$$

It results from another fundamental theorem in the theory of almost periodic functions* that we may approximate uniformly to $\phi_\epsilon(x)$ by a polynomial

$$\sum_1^N C_k e^{-i\Lambda_k x}. \tag{31.33}$$

Over $(-D, D)$ let

$$f(x) \sim \sum_{-\infty}^{\infty} d_n e^{in\pi x/D}. \tag{31.34}$$

Then within $(-D + 2\epsilon, D - 2\epsilon)$

$$f_\epsilon(x) = \sum_{-\infty}^{\infty} d_k \frac{4D^2}{n^2\pi^2\epsilon^2} \sin^2 \frac{n\pi\epsilon}{2D} e^{in\pi x/D}, \tag{31.35}$$

and

$$\phi_\epsilon(x) = \sum_{-\infty}^{\infty} d_k \frac{4D^3}{n^3\pi^3\epsilon^3} \sin^2 \frac{n\pi\epsilon}{2D} \sin \frac{n\pi\epsilon}{D} e^{in\pi x/D}. \tag{31.36}$$

Thus

$$\begin{aligned} &\lim_{\epsilon\to 0} \int_{-D+2\epsilon}^{D-2\epsilon} |f(x) - \phi_\epsilon(x)|^2 \, dx \\ &\leqq \lim_{\epsilon\to 0} 2D \sum_{n\to-\infty}^{\infty} \left| d_n \right|^2 \left| 1 - \frac{4D^3}{n^3\pi^3\epsilon^3} \sin^2 \frac{n\pi\epsilon}{2D} \sin \frac{n\pi\epsilon}{D} \right|^2 = 0. \end{aligned} \tag{31.37}$$

By an argument exactly like that used to establish (31.13), if $2D > A + 4\epsilon$,

$$\lim_{\epsilon\to 0} \int_x^{x+A} |f(x) - \phi_\epsilon(x)|^2 \, dx = 0 \tag{31.38}$$

* Ibid., p. 29.

uniformly in x over $(-\infty, \infty)$. It follows from (31.33) that we can find for any ϵ a polynomial $\sum_1^n l_k e^{-i\Lambda_k x}$ such that

$$(31.39) \qquad \int_x^{x+A} \left| f(x) - \sum_1^n l_k e^{-i\Lambda_k x} \right|^2 dx < \epsilon$$

uniformly over $(-\infty, \infty)$, hence,

$$(31.40) \qquad \int_x^{x+1} \left| f(x) - \sum_1^n l_k e^{-i\Lambda_k x} \right|^2 dx < \epsilon \left[\frac{1}{A} + 1 \right],$$

and Theorem XL is proved.

Let us notice that in a classical theorem of Stepanoff a function almost periodic in the sense Stepanoff 2 has associated with it a denumerable set $\{\lambda_n\}$ from which the exponents Λ_k of (31.40) may be selected. Furthermore, if we are given a pseudoperiodic function $f(x)$ with the exponents $\{\lambda_n\}$ we may write

$$(31.41) \qquad \lim_{T\to\infty} \frac{1}{2T} \int_{-T}^{T} | f(x) |^2 dx = \sum_1^\infty \left| \lim_{T\to\infty} \frac{1}{2T} \int_{-T}^{T} f(x) e^{-i\lambda_n x} dx \right|^2$$

and

$$(31.42) \quad \lim_{N\to\infty} \lim_{T_1\to\infty} \frac{1}{2T_1} \int_{-T_1}^{T_1} \left| f(x) - \sum_1^N \lim_{T\to\infty} \frac{e^{i\lambda_n x}}{2T} \int_{-T}^{T} f(\xi) e^{-i\lambda_n \xi} d\xi \right|^2 dx = 0 .$$

It follows from (31.41), (31.08) and the Riesz-Fischer theorem that over any range of length A, we have a function $g(x)$ such that uniformly in y,

$$(31.43) \qquad \lim_{N\to\infty} \int_y^{y+A} \left| g(x) - \sum_1^N \lim_{T\to\infty} \frac{e^{i\lambda_n x}}{2T} \int_{-T}^{T} f(\xi) e^{-i\lambda_n \xi} d\xi \right|^2 dx = 0 .$$

This function $g(x)$ will clearly be pseudoperiodic, as will $f(x) - g(x)$. By (31.42–3) and the Minkowski inequality,

$$(31.44) \qquad \lim_{T\to\infty} \frac{1}{2T} \int_{-T}^{T} | f(x) - g(x) |^2 dx = 0 ,$$

from which it will readily result that for all y,

$$(31.45) \qquad \int_y^{y+A} | f(x) - g(x) |^2 dx = 0 ,$$

or that $f(x)$ is equivalent to $g(x)$. Otherwise we should have some $\int_y^{y+A}$ not equivalent to 0, whence by pseudoperiodicity no $\int_y^{y+A}$ could be equivalent to 0, and (31.44) would be impossible. Thus

$$(31.46) \qquad \lim_{N\to\infty} \int_y^{y+A} \left| f(x) - \sum_1^N \lim_{T\to\infty} \frac{e^{i\lambda_n x}}{2T} \int_{-T}^{T} f(\xi) e^{-i\lambda_n \xi} d\xi \right|^2 dx = 0$$

uniformly in y.

In general, let $\sum_1^\infty |a_n|^2$ converge, and let (31.02) not vanish. It will follow, as in (31.43), that there exists a function $g(x)$ as in (31.43), which over any length A is the uniform limit in the mean of the functions $\sum_1^N a_n e^{i\lambda_n x}$ as $n \to \infty$. An easy procedure to the limit from (31.07) and (31.11) will show that if (31.02) exceeds L,

$$(31.47)\qquad \frac{9\pi^3}{4L}\sum_1^\infty |a_n|^2 \geqq \int_y^{y+9\pi/L} |g(x)|^2\,dx \geqq \frac{\pi}{9L}\sum_1^\infty |a_n|^2,$$

and from this it follows that $g(x)$ is pseudoperiodic.

It results at once from (31.47) that the convergence theory of the formal series $\sum_1^\infty a_n e^{i\lambda_n x}$ of pseudoperiodic functions follows closely the lines laid down in Theorem XXXVIII. Indeed, the methods of proof of this theorem allow us to conclude directly that

THEOREM XLI. *If $f(x)$ is a pseudoperiodic function with the formal series*

$$(31.475)\qquad \sum_1^\infty e^{i\lambda_n x} \lim_{T\to\infty} \frac{1}{2T}\int_{-T}^{T} f(\xi)\, e^{-i\lambda_n \xi}\,d\xi = \sum_1^\infty a_n e^{i\lambda_n x},$$

and if over $(-\pi, \pi)$

$$(31.48)\qquad f(x) \sim \frac{1}{2\pi}\sum_{-\infty}^{\infty} e^{inx}\int_{-\pi}^{\pi} f(\xi)\, e^{-in\xi}\,d\xi = \sum_{-\infty}^{\infty} b_n e^{in\xi},$$

then over $(-\pi+\epsilon, \pi-\epsilon)$ we have uniformly

$$(31.49)\qquad \lim_{N\to\infty}\left(\sum_{|\lambda_n|<N} a_n e^{i\lambda_n x} - \sum_{-N}^{N} b_n e^{inx}\right) = 0.$$

In particular, the convergence, Cesàro summability, etc., of the $e^{i\lambda_n x}$ series of $f(x)$ are the same as those of its Fourier series uniformly over $(-\pi+\epsilon, \pi-\epsilon)$.

32. Theorems on lacunary series.

We proceed to prove

THEOREM XLII. *Let no a_n vanish, and $\sum_{-\infty}^{\infty} |a_n|^2$ converge, and let $\cdots < \lambda_{-n} < \cdots < \lambda_{-1} < \lambda_0 < \lambda_1 < \cdots < \lambda_n < \cdots$. Let**

$$(32.01)\qquad \lim_{n\to\pm\infty} (\lambda_{n+1} - \lambda_n) = \infty.$$

Let

$$(32.02)\qquad f(x) = \underset{N\to\infty}{\text{l.i.m.}} \sum_{-N}^{N} a_n e^{i\lambda_n x}$$

* With the aid of the Fabry gap theorem (vide infra), we may prove the more general theorem derived from Theorem XLII by replacing (32.06) by

$$\lim_{n\to\pm\infty} \frac{\lambda_n}{n} = \infty.$$

over every finite range. Let $f(x)$ be equivalent to zero over any interval, (a, b). Then $f(x)$ is equivalent to zero over every interval, and all the a_n's vanish.

To begin with, let L be any positive quantity. Let N be so big that if $|n| > N$,

$$\lambda_{n+1} - \lambda_n > L. \tag{32.03}$$

Let $\epsilon < (b - a)/2$. Let $\psi(x)$, which can be constructed by orthogonalization, belong to L_2 over $(-\epsilon, \epsilon)$, and let $\int_{-\epsilon}^{\epsilon} \psi(x)\, e^{-inx}\, dx$ not vanish for a given λ_k, but vanish for all λ_n's (other than the given one, possibly) for which $|n| \leqq N$. Let

$$\begin{aligned} g(x) &= \int_{-\epsilon}^{\epsilon} \psi(\xi)\, f(x - \xi)\, d\xi \\ &= \int_{-\epsilon}^{\epsilon} \psi(\xi) \operatorname*{l.i.m.}_{N\to\infty} \sum_{-N}^{N} a_n\, e^{i\lambda_n(x-\xi)}\, d\xi \\ &= \sum_{-\infty}^{\infty} a_n \int_{-\epsilon}^{\epsilon} \psi(\xi)\, e^{-i\lambda_n\xi}\, e^{i\lambda_n x}\, d\xi. \end{aligned} \tag{32.04}$$

Then $g(x)$ will vanish over $(a + \epsilon, b - \epsilon)$ since $f(x)$ vanishes over (a, b), and will be the uniform limit of series (32.04), which will contain a term in $e^{i\lambda_k x}$ with a non-zero coefficient, but which will contain no two λ's closer together than L. Then by (31.11), $g(x)$ cannot be identically zero over any interval of length $9\pi/L$, and if $L > 9\pi/(b - a - 2\epsilon)$ this leads to a contradiction.

Let it be noted that Theorem XLII resembles Theorem XXXVI in being a gap theorem, but differs in that λ_n need not equal $-\lambda_{-n}$ and in that the λ_n's are not multiples of a common unit. On the other hand, Theorem XXXVI is concerned only with the lower density of the λ_n's.

A gap theorem closely related to Theorem XLII is

THEOREM XLII′. *Let $\cdots < \lambda_{-n} < \cdots < \lambda_{-1} < \lambda_0 < \lambda_1 < \cdots < \lambda_n < \cdots$, and let* (32.01) *hold. Let*

$$\sum_{-\infty}^{\infty} |a_n^2|\, r^{2|\lambda_n|} < \infty \tag{32.05}$$

for $0 < r < 1$. Let

$$f(r, x) = \operatorname*{l.i.m.}_{N\to\infty} \sum_{-N}^{N} a_n\, r^{|\lambda_n|}\, e^{i\lambda_n x} \tag{32.06}$$

over every finite range. Let us define

$$f(x) = \operatorname*{l.i.m.}_{r\to 1} f(r, x) \tag{32.07}$$

for any interval (a, b) over which the indicated limit in the mean exists. Then if $f(x)$ exists over any interval, it exists over every interval. If $f(x)$ is over any interval the n-fold integral of a function of L_2, it is such an integral over every interval, apart from the arbitrary set of measure zero over which it is indeterminate. If $f(x)$ is infinitely often differentiable over (a, b), if

$$\int_a^b |f^{(n)}(x)|^2 \, dx < D^n A_n^2, \tag{32.08}$$

where D is independent of n, and if

$$A_n > n!, \tag{32.09}$$

then if $d > c$, there exists a positive P, such that

$$\int_c^d |f^{(n)}(x)|^2 \, dx < P^n A_n^2. \tag{32.10}$$

If $f(x)$ is analytic over any interval, it is everywhere analytic.

This theorem has been proved by N. Wiener in a paper forthcoming in the Pisa Annali. It has as an immediate corollary the result that if in the power series $\Sigma a_n z^n$, the gaps between successive exponents of terms with non-null coefficients tend to infinity, the circle of convergence of the series is a natural boundary. Each arc, that is, must contain a singularity. Again, Theorem XLII is an immediate corollary, for if $f(x)$ vanishes over any interval, it is analytic everywhere, and hence must vanish everywhere.

Wiener has made an attempt to use methods of this type to prove the celebrated Fabry gap theorem.* This theorem yields the circle of convergence as a natural boundary on the lighter hypothesis, not that $\lambda_{n+1} - \lambda_n \to \infty$, but that $\lambda_n/n \to \infty$. So far he has had no success. On the other hand, the methods indicated are as applicable to Dirichlet as to Taylor series.

We now come to the proof of Theorem XLII′. Let us notice that after (a, b) is fixed, we may take

$$\text{l.u.b.}\, [\lambda_{n+1} - \lambda_n] \tag{32.11}$$

as large as we like, say $> L$, for to do so, we have only to remove from $f(r, x)$ a polynomial in $e^{i\lambda_k x}$, tending to a definite analytic limit as $r \to 1$. Such a function is analytic in x, and if we write it $P_r(x)$, there exists a constant K such that

$$|P_r^{(n)}(x)| < K^n. \tag{32.12}$$

Thus there is no restriction in putting

$$\lambda_{n+1} - \lambda_n > \frac{9\pi}{b-a} \qquad (n = 0, 1, 2, \cdots; -1, -2, \cdots). \tag{32.13}$$

* Cf. P. Dienes, *The Taylor Series*, Oxford, 1931, pp. 372 ff.

Thus by (31.11) and the following discussion, if

$$\left[\frac{d-c}{b-a}\right] = \nu, \tag{32.14}$$

we have for an appropriate constant B,

$$\int_c^d |f(r, x)|^2 dx \leqq B(\nu + 1) \int_a^b |f(r, x)|^2 dx, \tag{32.15}$$

and similarly

$$\int_c^d |f^{(n)}(r, x)|^2 dx \leqq B(\nu + 1) \int_a^b |f^{(n)}(r, x)|^2 dx. \tag{32.16}$$

A further result of the same type is

$$\int_c^d |f(r, x) - f(s, x)|^2 dx \leqq B(\nu + 1) \int_a^b |f(r, x) - f(s, x)|^2 dx, \tag{32.17}$$

and as the existence of $f(x)$ requires that

$$\lim_{r,s\to 1} \int_a^b |f(r, x) - f(s, x)|^2 dx = 0, \tag{32.18}$$

it follows that

$$\lim_{r,s\to 1} \int_c^d |f(r, x) - f(s, x)|^2 dx = 0. \tag{32.19}$$

Thus by the Riesz-Fischer theorem, $f(x)$ exists over (c, d). In exactly the same way, we may show that if $f(x)$ is the n-fold integral of the function $f^{(n)}(x)$ belonging to L_2 over (a, b), we may extend $f^{(n)}(x)$ to (c, d), and

$$\int_c^d |f^{(n)}(x)|^2 dx \leqq B(\nu + 1) \int_a^b |f^{(n)}(x)|^2 dx. \tag{32.20}$$

This enables us to make a simple transition from (32.08) to (32.10).

If $f(x)$ is analytic over (a, b), then over some interval $(a + \epsilon, b - \epsilon)$, we have*

$$|f^{(n)}(x)| < c^n n!, \tag{32.21}$$

and vice versa. Let (c, d) be any interval containing (a, b). We have

$$\int_a^b |f^{(n)}(x)|^2 dx < c_0^{2n} (n!)^2. \tag{32.22}$$

* For in such an instance, $\Sigma f^{(n)}(x)\,(\xi - n)^n/n!$ has a positive radius of convergence for every x, and hence, by the Heine-Borel theorem, this radius of convergence has a lower bound.

Hence for some other c_0

$$\int_c^d | f^{(n)}(x) |^2 \, dx < c_0^{2n} (n!)^2. \tag{32.23}$$

Thus over (c, d)

$$\begin{aligned} | f^{(n-1)}(x) | &\leqq | f^{(n-1)}(a) | + \left| \int_a^x f^{(n)}(\xi) \, d\xi \right| \\ &\leqq c_0^n \, n! + \left\{ | x - a | \int_a^x | f^{(n)}(\xi) |^2 \, d\xi \right\}^{1/2} \\ &\leqq c_1^n \, n! + (c_2 \cdot c_3^n (n!)^2)^{1/2} \\ &\leqq c_4^n \, n!. \end{aligned} \tag{32.24}$$

It follows that

$$| f^{(n)}(x) | \leqq c_5^n \, n!. \tag{32.25}$$

However, this implies the statement that $f(x)$ is analytic over every $(c + \epsilon, d - \epsilon)$.

It is interesting to note the complete picture formed by the results of the present and the last chapter, particularly in what concerns the closure properties of $\{e^{i\lambda_n x}\}$ where $| \lambda_n - n | < L$. We have already seen that if $\lambda_{-n} = -\lambda_n$ the set becomes closed on the adjunction at most of a finite number of terms, and ceases to be closed if we drop at most a finite number (Theorem XXXIII). If we consider the same set over any interval of length greater than $4\pi/L$ it results at once from Theorem XL that the set is one in which 0 cannot be expanded with coefficients, the sum of the squares of which converges. On the other hand, over an interval $(-1/(\pi L), 1/(\pi L))$, the set (Theorem XXXVIII) contains a subset with an L_2 expansion theory substantially the same as that for expansions in ordinary Fourier series.

In general, the results of the present chapter, although they go much farther than any previous researches in this direction, are fragmentary at almost every turn. In scarcely any case do our theorems state necessary and sufficient conditions, and as a matter of fact it is highly improbable that most of our results are of this nature. Our inequalities are nowhere shown to be the best possible. It is the opinion of the authors that these limitations are inherent in the methods employed; that while their power has not been completely exhausted, some radically new idea is needed if we are to bring this theory into its definitive shape.

CHAPTER VIII

GENERALIZED HARMONIC ANALYSIS IN THE COMPLEX DOMAIN

33. **Relevant theorems of generalized harmonic analysis.** Wiener* and others have developed a theory of generalized harmonic analysis. This is a theory of developments in series of trigonometric functions which includes as particular cases the Fourier series and the Fourier integral, but which also includes theories such as that of white light, which do not come under either of the above headings. His theory has so far been confined to functions which may be complex-valued but are of real arguments. Now every theory of the harmonic analysis of functions of arguments in the real domain has an associated theory of functions of arguments in the complex domain. In the case of the Fourier integral, the associated theory is that of the Laplace integral; in the case of the Fourier series, the associated theory is that of the Taylor and Laurent series; and in the case of non-harmonic developments in discrete trigonometric functions such as are found in Bohr's theory of almost periodic functions, the associated theory is that of the Dirichlet theory. It is the purpose of this chapter to extend the Wiener theory in the same sense and to subsume under this general theory certain theorems concerning almost periodic functions. To this end we shall have to carry over certain results from Wiener's book. It does not come under the purpose of this chapter to prove these theorems or, indeed, to give any very detailed discussion of their significance. We shall simply remark that $f(x)$ represents the function subject to harmonic analysis; that $s(u)$ represents the integral of the Fourier transform (which itself does not exist) up to the argument u, and that $S(u)$ represents the total energy in the spectrum of f up to the frequency used. $\phi(x)$, as we shall define it below, is the so-called Faltung of f with itself, and $S(u)$ may be determined in terms of ϕ alone. S is the class of functions $f(x)$ determining a "spectrum" $S(u)$, and S' is the sub-class of S for which in a certain sense the spectrum contains no energy at infinite frequency.

The theorems and definitions which we shall quote come from Chapters III and IV of Wiener's book. They read as follows:

THEOREM 20. *Let $f(x)$ be a measurable function for which*

(33.01)
$$\frac{1}{2T}\int_{-T}^{T} | f(x) |^2 \, dx$$

is bounded in T. Then

(33.02)
$$\int_{-\infty}^{\infty} \frac{| f(x) |^2}{1 + x^2} \, dx < \infty .$$

* *The Fourier Integral and Certain of its Applications*, Cambridge, 1933.

DEFINITION.

$$s(u) = \underset{A\to\infty}{\text{l.i.m.}} \frac{1}{(2\pi)^{1/2}}\left[\int_1^A + \int_{-A}^{-1}\right]\frac{f(x)\,e^{-iux}}{-ix}\,dx \tag{33.03}$$
$$+ \frac{1}{(2\pi)^{1/2}}\int_{-1}^{1} f(x)\,\frac{e^{-iux}-1}{-ix}\,dx\,.$$

Let it be noted that under the hypothesis of Theorem 20, $s(u)$ will exist for almost every u.

THEOREM 22. *Under the hypothesis of Theorem* 20,

$$\lim_{\epsilon\to 0}\frac{1}{4\pi\epsilon}\int_{-\infty}^{\infty} |\, s(u+\epsilon) - s(u-\epsilon)\,|^2\,du = \lim_{T\to\infty}\frac{1}{2T}\int_{-T}^{T} |\, f(x)\,|^2\,dx \tag{33.04}$$

in the sense that if either side exists, the other exists and has the same value.

DEFINITION.

$$\phi(x) = \lim_{T\to\infty}\frac{1}{2T}\int_{-T}^{T} f(x+\xi)\,\bar{f}(\xi)\,d\xi\,. \tag{33.05}$$

DEFINITION. *S is the class of measurable functions $f(x)$ for which $\phi(x)$ exists for every real x. S' is the class of functions of S for which $\phi(x)$ is continuous.*

THEOREM 36. *If $f(x)$ belongs to S, then*

$$S(u) = \frac{1}{(2\pi)^{1/2}}\int_{-\infty}^{\infty}\phi(x)\,\frac{e^{-iux}-1}{-ix}\,dx \tag{33.06}$$

exists for every u.

On page 162 of Wiener's book there is a formula which reads

$$\sigma(u) = \text{const.} + \underset{\epsilon\to 0}{\text{l.i.m.}}\frac{1}{2\epsilon(2\pi)^{1/2}}\int_0^u |\, s(u+\epsilon) - s(u-\epsilon)\,|^2\,du\,, \tag{33.07}$$

and is numbered (21.257). This formula states more than has actually been established at the point in question, and should read

$$\sigma(u) = \underset{\epsilon\to 0}{\text{l.i.m.}}\left[\psi(\epsilon) + \frac{1}{2\epsilon(2\pi)^{1/2}}\int_0^u |\, s(u+\epsilon) - s(u-\epsilon)\,|^2\,du\right]. \tag{33.08}$$

In Theorem 36, Wiener shows that if $f(x)$ belongs to S, $\sigma(u) - S(u)$ is a constant, except at most over a null set. Thus except at most over a null set, we have

$$S(u) = \underset{\epsilon\to 0}{\text{l.i.m.}}\left[\psi(\epsilon) + \frac{1}{2\epsilon(2\pi)^{1/2}}\int_0^u |\, s(u+\epsilon) - s(u-\epsilon)\,|^2\,du\right]. \tag{33.09}$$

THEOREM 30. *Let $f(x)$ belong to S. Let $xK(x)$ belong to L_1, and $(1+|\,x\,|)K(x)$ to L_2. Let*

$$g(x) = \int_{-\infty}^{\infty} K(x-\xi)\,f(\xi)\,d\xi\,. \tag{33.10}$$

Let $S(u)$ be defined as in (33.06), *and let*

(33.11)
$$T(u) = \frac{1}{(2\pi)^{1/2}} \int_{-\infty}^{\infty} \frac{(e^{-iux} - 1)\, dx}{-ix} \lim_{T\to\infty} \frac{1}{2T} \int_{-T}^{T} dw \times \int_{-\infty}^{\infty} K(x + w - \xi) f(\xi)\, d\xi \int_{-\infty}^{\infty} \overline{K(w - \xi)}\, \overline{f(\xi)}\, d\xi .$$

Then

(33.12)
$$T(u) = \int_0^u \left| \int_{-\infty}^{\infty} K(\xi)\, e^{-iu\xi}\, d\xi \right|^2 dS(u) .$$

Here we replace the σ of Wiener's formula by S.

Lemma 29_3. *Under the hypothesis of Theorem* 30,

(33.13)
$$\lim_{\epsilon\to 0} \frac{1}{\epsilon} \int_{-\infty}^{\infty} \left| t(u+\epsilon) - t(u-\epsilon) - \{s(u+\epsilon) - s(u-\epsilon)\} \times \int_{-\infty}^{\infty} K(\xi)\, e^{-iu\xi}\, d\xi \right|^2 du = 0 ,$$

where

(33.14)
$$t(u) = \underset{A\to\infty}{\text{l.i.m.}} \frac{1}{(2\pi)^{1/2}} \left[\int_1^A + \int_{-A}^{-1} \right] \frac{g(x)\, e^{-iux}\, dx}{-ix} + \frac{1}{(2\pi)^{1/2}} \int_{-1}^{1} \frac{g(x)(e^{-iux} - 1)\, dx}{-ix} .$$

Lemma 29_6. *Under the hypothesis of Theorem* 30, *$g(x)$ will belong to S'.*

34. Cauchy's theorem. Let

(34.01)
$$\int_{-A}^{A} | f(x) |^2\, dx = O(A) .$$

Then, as in (33.02),

(34.02)
$$\int_{-A}^{A} \frac{| f(x) |^2\, dx}{1 + x^2} = O(1) .$$

Thus if $f(x + i\eta)$ is analytic over $a \leqq x \leqq b$ and

(34.03)
$$\int_{-A}^{A} | f(x + iy) |^2\, dy = O(A)$$

uniformly in x over $a \leqq x \leqq b$, the function

(34.04)
$$\frac{f(x + iy)}{x + iy - c} \qquad (c > b > a)$$

belongs uniformly to L_2 over $a \leqq x \leqq b$. Thus by (3.38), if $a < x < b$,

$$\frac{f(x+iy)}{x+iy-c} = \frac{1}{2\pi}\int_{-\infty}^{\infty} \frac{f(b+i\eta)\,d\eta}{(b+i\eta-c)(b+i\eta-x-iy)} - \frac{1}{2\pi}\int_{-\infty}^{\infty} \frac{f(a+i\eta)\,d\eta}{(a+i\eta-c)(a+i\eta-x-iy)}. \tag{34.05}$$

Now,

$$\frac{x+iy-c}{(b+i\eta-c)(b+i\eta-x-iy)} = \frac{1}{b+i\eta-x-iy} - \frac{1}{b+i\eta-c}, \tag{34.06}$$

so that

$$f(x+iy) = \frac{1}{2\pi}\int_{-\infty}^{\infty} f(b+i\eta)\,d\eta\left(\frac{1}{b+i\eta-x-iy} - \frac{1}{b+i\eta-c}\right) - \frac{1}{2\pi}\int_{-\infty}^{\infty} f(a+i\eta)\,d\eta\left(\frac{1}{a+i\eta-x-iy} - \frac{1}{a+i\eta-c}\right). \tag{34.07}$$

Let us form

$$\begin{gathered} \phi_n(x) = 0 \quad (x < n); \qquad \phi_n(x) = \frac{(x-n)^2}{2} \quad (n \leqq x < n+1); \\ \phi_n(x) = 1 - \frac{(n+2-x)^2}{2} \quad (n+1 \leqq x < n+2); \\ \phi_n(x) = 1 \quad (n+2 \leqq x); \end{gathered} \tag{34.08}$$

and the transforms

$$K_1(z) = \int_{-\infty}^{\infty} \phi_n(\xi)\, e^{\xi z}\, d\xi \tag{34.09}$$

and

$$K_2(z) = \int_{-\infty}^{\infty} (\phi_n(\xi) - 1)\, e^{\xi z}\, d\xi. \tag{34.10}$$

We shall have at infinity for any $x < 0$,

$$K_1(x+iy) = O\left(\frac{1}{y^3}\right), \tag{34.11}$$

and for any $x > 0$,

$$K_2(x+iy) = O\left(\frac{1}{y^3}\right). \tag{34.12}$$

Now let us consider the function

$$h_1(z) = \frac{1}{z - x - iy} - \frac{1}{z - c} - K_1(x + iy - z) \qquad (c > \Re(z) > x)$$

$$(34.13) \qquad = \int_0^n e^{\xi(x+iy-z)}\, d\xi + \int_n^{n+1} \left(1 - \frac{(\xi - n)^2}{2}\right) e^{\xi(x+iy-z)}\, d\xi + \int_{n+1}^{n+2} \frac{(n + 2 - \xi)^2}{2} e^{\xi(x+iy-z)}\, d\xi + \int_0^\infty e^{\xi(z-c)}\, d\xi\,,$$

and

$$h_2(z) = \frac{1}{z - x - iy} - \frac{1}{z - c} - K_2(x + iy - z) \qquad (\Re(z) < x)$$

$$(34.14) \qquad = \int_0^n e^{\xi(x+iy-z)}\, d\xi + \int_n^{n+1} \left(1 - \frac{(\xi - n)^2}{2}\right) e^{\xi(x+iy-z)}\, d\xi + \int_{n+1}^{n+2} \frac{(n + 2 - \xi)^2}{2} e^{\xi(x+iy-z)}\, d\xi + \int_0^\infty e^{\xi(z-c)}\, d\xi\,.$$

We see that $h_1(z)$ and $h_2(z)$ represent the same analytic function, which is moreover $O(1/\Im(z)^2)$ at infinity. Thus by Cauchy's theorem,

$$(34.15) \qquad 0 = \frac{1}{2\pi} \int_{-\infty}^\infty f(b + i\eta)\, h_1(b + i\eta)\, d\eta - \frac{1}{2\pi} \int_{-\infty}^\infty f(a + i\eta)\, h_2(a + i\eta)\, d\eta\,,$$

and by (34.07)

$$(34.16) \qquad f(x + iy) = \frac{1}{2\pi} \int_{-\infty}^\infty f(b + i\eta)\, K_1(x + iy - b - i\eta)\, d\eta - \frac{1}{2\pi} \int_{-\infty}^\infty f(a + i\eta)\, K_2(x + iy - a - i\eta)\, d\eta\,.$$

By (34.02) these integrals are absolutely convergent.

Let us now assume that $f(a + iy)$ and $f(b + iy)$ both belong to S, and that for $a \leqq x \leqq b$, (34.03) is uniformly satisfied. Let us put

$$(34.17) \qquad s(x, u) = \underset{A\to\infty}{\text{l.i.m.}}\, \frac{1}{(2\pi)^{1/2}} \left[\int_1^A + \int_{-A}^{-1}\right] \frac{f(x + iy)\, e^{-iuy}\, dy}{-iy} + \frac{1}{(2\pi)^{1/2}} \int_{-1}^1 \frac{f(x + iy)\, (e^{-iuy} - 1)\, dy}{-iy}\,.$$

Then by (33.13) and (34.16),

$$(34.18) \qquad \lim_{\epsilon\to 0} \frac{1}{\epsilon} \int_{-\infty}^\infty \big| s(x, u + \epsilon) - s(x, u - \epsilon) - \{s(b, u + \epsilon) - s(b, u - \epsilon)\} \times \phi_n(u)\, e^{u(x-b)} + \{s(a, u + \epsilon) - s(a, u - \epsilon)\}\, (\phi_n(u) - 1)\, e^{u(x-a)} \big|^2\, du = 0\,,$$

where

$$\phi_n(u)\, e^{u(x-b)} = \frac{1}{2\pi} \int_{-\infty}^{\infty} K_1(x - b + iy)\, e^{-iuy}\, dy\,, \tag{34.19}$$

and

$$(\phi_n(u) - 1)\, e^{u(x-a)} = \frac{1}{2\pi} \int_{-\infty}^{\infty} K_2(x - a + iy)\, e^{-iuy}\, dy\,. \tag{34.20}$$

It will immediately follow by proper choice of n that over any finite range,

$$\begin{aligned} &\lim_{\epsilon\to 0} \frac{1}{\epsilon} \int_A^B |\, s(x, u+\epsilon) - s(x, u-\epsilon) - \{s(b, u+\epsilon) - s(b, u-\epsilon)\} e^{u(x-b)} \,|^2\, du \\ &\qquad = \lim_{\epsilon\to 0} \frac{1}{\epsilon} \int_A^B |\, s(x, u+\epsilon) - s(x, u-\epsilon) \\ &\qquad\qquad - \{s(a, u+\epsilon) - s(a, u-\epsilon)\} e^{u(x-a)} \,|^2\, du \\ &\qquad = 0\,. \end{aligned} \tag{34.21}$$

Furthermore, if ϵ is smaller than some quantity independent of B and x, using (33.04),

$$\begin{aligned} &\frac{1}{\epsilon} \int_B^{\infty} |\, \{s(b, u+\epsilon) - s(b, u-\epsilon)\}\, e^{u(x-b)} \,|^2\, du \\ &\qquad \leqq 2e^{2B(x-b)} \frac{1}{2\epsilon} \int_{-\infty}^{\infty} |\, s(b, u+\epsilon) - s(b, u-\epsilon) \,|^2\, du \\ &\qquad \leqq \text{const.}\ e^{2B(x-b)}\,. \end{aligned} \tag{34.22}$$

Similarly,

$$\frac{1}{\epsilon} \int_{-\infty}^{A} |\, \{s(a, u+\epsilon) - s(a, u-\epsilon)\}\, e^{u(x-a)} \,|^2\, du \leqq \text{const.}\ e^{2A(x-a)}\,. \tag{34.23}$$

Thus

$$\begin{aligned} &\lim_{\epsilon\to 0} \frac{1}{\epsilon} \int_{-\infty}^{\infty} |\, s(x, u+\epsilon) - s(x, u-\epsilon) \,|^2\, du \\ &\qquad = \lim_{A\to\infty} \lim_{\epsilon\to 0} \frac{1}{\epsilon} \int_{-A}^{A} |\, s(b, u+\epsilon) - s(b, u-\epsilon) \,|^2\, e^{2u(x-b)}\, du \\ &\qquad = \lim_{A\to\infty} \lim_{\epsilon\to 0} \frac{1}{\epsilon} \int_{-A}^{A} |\, s(a, u+\epsilon) - s(a, u-\epsilon) \,|^2\, e^{2u(x-a)}\, du\,, \end{aligned} \tag{34.24}$$

in case

$$\lim_{\epsilon\to 0} \frac{1}{\epsilon} \int_{-A}^{A} |\, s(b, u+\epsilon) - s(b, u-\epsilon) \,|^2\, e^{2u(x-b)}\, du \tag{34.25}$$

and

$$\lim_{\epsilon \to 0} \frac{1}{\epsilon} \int_{-A}^{A} | s(a, u + \epsilon) - s(a, u - \epsilon) |^2 e^{2u(x-a)} \, du \tag{34.26}$$

exist for all A's of an increasing sequence tending to infinity.

Let us put

$$\begin{aligned} S(x, u) &= \frac{1}{(2\pi)^{1/2}} \int_{-\infty}^{\infty} \frac{(e^{-iu\xi} - 1) \, d\xi}{-i\xi} \\ &\quad \times \lim_{T \to \infty} \frac{1}{2T} \int_{-T}^{T} f(x + iy + i\xi) \overline{f(x + iy)} \, dy ; \end{aligned} \tag{34.27}$$

$$\begin{aligned} T_1(x, u) &= \frac{1}{(2\pi)^{1/2}} \int_{-\infty}^{\infty} \frac{(e^{-iu\xi} - 1) \, d\xi}{- i\xi} \lim_{T \to \infty} \frac{1}{2T} \int_{-T}^{T} \frac{dy}{4\pi^2} \\ &\quad \times \int_{-\infty}^{\infty} f(b + i\eta) \, K_1(x + iy + i\xi - b - i\eta) d\eta \\ &\quad \times \int_{-\infty}^{\infty} \overline{f(b + i\eta)} \, \overline{K_1(x + iy - b - i\eta)} \, d\eta ; \end{aligned} \tag{34.28}$$

$$\begin{aligned} T_2(x, u) &= \frac{1}{(2\pi)^{1/2}} \int_{-\infty}^{\infty} \frac{(e^{-iu\xi} - 1) \, d\xi}{- i\xi} \lim_{T \to \infty} \frac{1}{2T} \int_{-T}^{T} \frac{dy}{4\pi^2} \\ &\quad \times \int_{-\infty}^{\infty} f(a + i\eta) \, K_2(x + iy + i\xi - a - i\eta) d\eta \\ &\quad \times \int_{-\infty}^{\infty} \overline{f(a + i\eta)} \, \overline{K_2(x + iy - a - i\eta)} \, d\eta . \end{aligned} \tag{34.29}$$

By Theorem 36 and Lemma 29_6, $T_1(x, u)$ and $T_2(x, u)$ will exist for every u. By (33.09) and (33.13),

$$\begin{aligned} T_1(x, u) &= \underset{\epsilon \to 0}{\text{l.i.m.}} \Bigg[\psi(\epsilon) + \frac{1}{2\epsilon \, (2\pi)^{1/2}} \\ &\quad \times \int_0^u | s(b, u + \epsilon) - s(b, u - \epsilon) |^2 \, [\phi_n(u)]^2 \, e^{2u(x-b)} \, du \Bigg] . \end{aligned} \tag{34.30}$$

Now, if we have a sequence of monotone functions $f_n(x)$ converging in the mean to a limit $f(x)$, we shall show that it converges almost everywhere to $f(x)$. We have uniformly

$$\frac{1}{2\epsilon} \int_{x-\epsilon}^{x+\epsilon} f(\xi) \, d\xi = \lim_{n \to \infty} \frac{1}{2\epsilon} \int_{x-\epsilon}^{x+\epsilon} f_n(\xi) \, d\xi . \tag{34.31}$$

From this it follows that independently of $\epsilon > 0$,

$$\underline{\lim}_{n \to \infty} f_n(x + \epsilon) \geqq \frac{1}{2\epsilon} \int_{x-\epsilon}^{x+\epsilon} f(\xi) \, d\xi , \qquad \overline{\lim}_{n \to \infty} f_n(x - \epsilon) \leqq \frac{1}{2\epsilon} \int_{x-\epsilon}^{x+\epsilon} f(\xi) \, d\xi . \tag{34.32}$$

Furthermore, by Weyl's lemma to the Riesz-Fischer theorem, there exists a subsequence $\{f_{n_k}(x)\}$ $(k = 1, 2, \cdots)$ of $\{f_n(x)\}$, converging almost everywhere to $f(x)$. Thus $f(x)$ is monotone over a set of points Σ differing from the whole line at most by a null set. This set is of course everywhere dense. Let $F(x)$ stand for the value of $f(x)$ thus defined for values of x in Σ. If y does not lie in Σ, let us put

$$f(y) = (1/2)\left[\underset{x \text{ in } \Sigma,\, x<y}{\text{l.u.b.}} F(x) + \underset{x \text{ in } \Sigma,\, x>y}{\text{g.l.b.}} F(x)\right] \tag{34.321}$$

and otherwise let $f(x) = F(x)$. Then $f(x)$ will be everywhere defined and monotone.

Again, by (34.32),

$$\begin{aligned} \underline{\lim}_{n\to\infty} f_n(x+\epsilon) &\geqq \overline{\lim}_{\eta\to 0} \frac{1}{2\eta}\int_{x-\eta}^{x+\eta} f(\xi)\,d\xi\,, \\ \overline{\lim}_{n\to\infty} f_n(x-\epsilon) &\leqq \underline{\lim}_{\eta\to 0} \frac{1}{2\eta}\int_{x-\eta}^{x+\eta} f(\xi)\,d\xi\,, \end{aligned} \tag{34.322}$$

and by the fundamental theorem of the calculus, we have almost everywhere

$$\underline{\lim}_{n\to\infty} f_n(x+\epsilon) \geqq f(x) \geqq \overline{\lim}_{n\to\infty} f_n(x-\epsilon)\,. \tag{34.323}$$

Thus almost everywhere

$$f(x-\epsilon) \leqq \underline{\lim}_{n\to\infty} f_n(x) \leqq \overline{\lim}_{n\to\infty} f_n(x) \leqq f(x+\epsilon) \tag{34.324}$$

and hence almost everywhere

$$\lim_{\epsilon\to 0} f(x-\epsilon) \leqq \underline{\lim}_{n\to\infty} f_n(x) \leqq \overline{\lim}_{n\to\infty} f_n(x) \leqq \lim_{\epsilon\to 0} f(x+\epsilon)\,. \tag{34.325}$$

Since a monotone function is almost everywhere continuous, this yields almost everywhere

$$f(x) \leqq \underline{\lim}_{n\to\infty} f_n(x) \leqq \overline{\lim}_{n\to\infty} f_n(x) \leqq f(x)\,. \tag{34.326}$$

Thus it results that for almost all x,

$$f(x) = \lim_{n\to\infty} f_n(x)\,. \tag{34.33}$$

This theorem allows us to use lim in place of l.i.m. in (34.30).

Now let n be taken as an arbitrarily large negative number in (34.30). Then for almost all A,

$$\begin{aligned} \lim_{\epsilon\to 0} \frac{1}{2\epsilon\,(2\pi)^{1/2}} \int_{-A}^{A} | \, s(b, u+\epsilon) - s(b, u-\epsilon) \, |^2 \, e^{2u(x-b)}\,du \\ = T_1(x, A) - T_1(x, -A)\,. \end{aligned} \tag{34.34}$$

Similarly, for a large positive choice of n, we get

$$\lim_{\epsilon\to 0} \frac{1}{2\epsilon\,(2\pi)^{1/2}} \int_{-A}^{A} |\, s(a, u+\epsilon) - s(a, u-\epsilon)\,|^2\, e^{2u(x-a)}\, du \tag{34.35}$$
$$= T_2(x, A) - T_2(x, -A).$$

Thus by formula (34.24),

$$\lim_{\epsilon\to 0} \frac{1}{\epsilon} \int_{-\infty}^{\infty} |\, s(x, u+\epsilon) - s(x, u-\epsilon)\,|^2\, du \tag{34.36}$$

exists for $a < x < b$, and has the value indicated in that formula.

A proof of exactly the same sort, somewhat more fussy in detail, but not differing at all in principle, will show that if $a < x < b$, and ξ is real,

$$\lim_{\epsilon\to 0} \frac{1}{\epsilon} \int_{-\infty}^{\infty} |\, s(x, u+\epsilon) - s(x, u-\epsilon)\,|^2\, e^{-iu\xi}\, du \tag{34.361}$$

exists, and equals

$$\lim_{A\to\infty} \lim_{\epsilon\to 0} \frac{1}{\epsilon} \int_{-A}^{A} |\, s(b, u+\epsilon) - s(b, u-\epsilon)\,|^2\, e^{2u(x-b)-iu\xi}\, du. \tag{34.362}$$

It is clear that

$$\overline{\lim_{A\to\infty}}\ \overline{\lim_{\epsilon\to 0}} \frac{1}{\epsilon} \left| \left[\int_{A}^{\infty} + \int_{-\infty}^{-A} \right] |\, s(b, u+\epsilon) - s(b, u-\epsilon)\,|^2\, e^{2u(x-b)-iu\xi}\, du \right| \tag{34.363}$$
$$\leqq \lim_{A\to\infty} \lim_{\epsilon\to 0} \frac{1}{\epsilon} \left[\int_{A}^{\infty} + \int_{-\infty}^{-A} \right] |\, s(b, u+\epsilon) - s(b, u-\epsilon)\,|^2\, e^{2u(x-b)}\, du,$$

and we have just proved in (34.34) that this is zero. Hence (34.362) exists, and we have only to prove its identity with (34.361). This will readily follow if for a set of values of A increasing to infinity,

$$\lim_{\epsilon\to 0} \frac{1}{\epsilon} \int_{-A}^{A} e^{-iu\xi} \{\, |\, s(x, u+\epsilon) - s(x, u-\epsilon)\,|^2 \tag{34.364}$$
$$- e^{2u(x-b)}\, |\, s(b, u+\epsilon) - s(b, u-\epsilon)\,|^2 \}\, du = 0.$$

By the use of the Schwarz inequality and the boundedness of

$$\frac{1}{\epsilon} \int_{-A}^{A} |\, s(x, u+\epsilon) - s(x, u-\epsilon)\,|^2\, du \quad \text{and} \tag{34.365}$$
$$\frac{1}{\epsilon} \int_{-A}^{A} e^{2u(x-b)}\, |\, s(b, u+\epsilon) - s(b, u-\epsilon)\,|^2\, du,$$

(34.364) may be carried back to

$$\lim_{\epsilon\to 0} \frac{1}{\epsilon} \int_{-A}^{A} |\, s(x, u+\epsilon) - s(x, u-\epsilon) \tag{34.366}$$
$$- e^{u(x-b)}\, (s(b, u+\epsilon) - s(b, u-\epsilon))\,|^2\, du = 0,$$

which goes back to (34.21).

Now let us define

$$
\begin{aligned}
s_\xi(x, u) &= \frac{1}{(2\pi)^{1/2}}\left[\int_1^A + \int_{-A}^{-1}\right]\frac{f(x+iy-i\xi)}{-iy}\, e^{-iuy}\, dy \\
&\quad + \frac{1}{(2\pi)^{1/2}}\int_{-1}^{1} f(x+iy-i\xi)\,\frac{e^{-iuy}-1}{-iy}\, dy\,.
\end{aligned} \tag{34.37}
$$

This will exist for the same reason as (34.17). As on page 158 of Wiener's book,

$$
\int_{-\infty}^{\infty} |\, s_\xi(x, u+\epsilon) - s_\xi(x, u-\epsilon) - e^{-iu\xi}(s(x, u+\epsilon) - s(x, u-\epsilon))\,|^2\, du = O(\epsilon^2)\,. \tag{34.38}
$$

As in that argument, this leads to

$$
\begin{aligned}
&\lim_{B\to\infty}\frac{1}{2B}\int_{-B}^{B} |\, f(x+iy-i\xi) + w\, f(x+iy)\,|^2\, dy \\
&\qquad = \lim_{\epsilon\to 0}\frac{1}{4\pi\epsilon}\int_{-\infty}^{\infty}(2 + w\, e^{iu\xi} + \bar{w}\, e^{-iu\xi})\,|\, s(u+\epsilon) - s(u-\epsilon)\,|^2\, du\,.
\end{aligned} \tag{34.39}
$$

If we take w successively to equal ± 1, $\pm i$, and add four such expressions as (34.39), with coefficients ± 1, $\pm i$, we get

$$
\begin{aligned}
&\lim_{B\to\infty}\frac{1}{2B}\int_{-B}^{B} f(x+iy-i\xi)\,\bar{f}(x+iy)\, dy \\
&\qquad = \lim_{\epsilon\to 0}\frac{1}{4\pi\epsilon}\int_{-\infty}^{\infty} |\, s(x, u+\epsilon) - s(x, u-\epsilon)\,|^2\, e^{-iu\xi}\, du\,,
\end{aligned} \tag{34.40}
$$

as in (21.17) of Wiener's book. Thus $f(x+iy)$ belongs to S over $a < x < b$.

By (34.40) we see that

$$
\begin{aligned}
&\lim_{\xi\to 0}\lim_{B\to\infty}\frac{1}{2B}\int_{-B}^{B} f(x+iy-i\xi)\,\bar{f}(x+iy)\, dy - \lim_{B\to\infty}\frac{1}{2B}\int_{-B}^{B} |\, f(x+iy)\,|^2\, dy \\
&= \lim_{\xi\to 0}\lim_{\epsilon\to 0}\frac{1}{4\pi\epsilon}\int_{-\infty}^{\infty}(1 - e^{-iu\xi})\,|\, s(x, u+\epsilon) - s(x, u-\epsilon)\,|^2\, du \\
&\leqq \overline{\lim_{\xi\to 0}}\;\overline{\lim_{\epsilon\to 0}}\;\frac{1}{4\pi\epsilon}\int_{-A}^{A} |\, 1 - e^{-iu\xi}\,|\;|\, s(x, u+\epsilon) - s(x, u-\epsilon)\,|^2\, du \\
&\quad + \overline{\lim_{\xi\to 0}}\;\overline{\lim_{\epsilon\to 0}}\;\frac{1}{4\pi\epsilon}\left[\int_A^{\infty} + \int_{-\infty}^{-A}\right] |\, 1 - e^{-iu\xi}\,|\;|\, s(x, u+\epsilon) - s(x, u-\epsilon)\,|^2\, du \\
&\leqq \overline{\lim_{\xi\to 0}}\;|\, 1 - e^{-iA\xi}\,|\;\lim_{\epsilon\to 0}\frac{1}{4\pi\epsilon}\int_{-\infty}^{\infty} |\, s(x, u+\epsilon) - s(x, u-\epsilon)\,|^2\, du \\
&\quad + \overline{\lim_{\epsilon\to 0}}\;\frac{1}{2\pi\epsilon}\left[\int_A^{\infty} + \int_{-\infty}^{-A}\right] |\, s(x, u+\epsilon) - s(x, u-\epsilon)\,|^2\, du \\
&\leqq \lim_{\epsilon\to 0}\frac{1}{2\pi\epsilon}\int_{-\infty}^{\infty} |\, s(x, u+\epsilon) - s(x, u-\epsilon)\,|^2\, du
\end{aligned} \tag{34.41}
$$

$$- \lim_{\epsilon\to 0} \frac{1}{2\pi\epsilon}\int_{-A}^{A} | s(x, u+\epsilon) - s(x, u-\epsilon) |^2 \, du .$$

We now appeal to (34.24), (34.21), and (34.34), and obtain in place of the last line of (34.41),

$$\begin{aligned}(34.42)\qquad &\lim_{B\to\infty}\lim_{\epsilon\to 0} \frac{1}{2\pi\epsilon}\int_{-B}^{B} | s(b, u+\epsilon) - s(b, u-\epsilon) |^2 e^{2u(x-b)} \, du \\ &\qquad - \lim_{\epsilon\to 0} \frac{1}{2\pi\epsilon}\int_{-A}^{A} | s(b, u+\epsilon) - s(b, u-\epsilon) |^2 e^{2u(x-b)} \, du ,\end{aligned}$$

which may be made as small as we please by making A large enough. Thus within $a < x < b$,

$$\lim_{B\to\infty} \frac{1}{2B}\int_{-B}^{B} f(x+iy+i\xi)\, \bar{f}(x+iy)\, dy$$

is continuous in ξ, and $f(x+iy)$ belongs to S' as a function of y.

35. **Almost periodic functions.** Let us return to (34.16). Let $f(a+iy)$ and $f(b+iy)$ be almost periodic in y: that is, let it be possible, whenever $\epsilon > 0$, to find trigonometrical polynomials

$$(35.01)\qquad P_1(y) = \sum_1^n A_k e^{i\Lambda_k y}$$

and

$$(35.02)\qquad P_2(y) = \sum_1^n B_k e^{iM_k y},$$

such that

$$(35.03)\qquad | f(a+iy) - P_1(y) | < \epsilon \qquad (-\infty < y < \infty),$$

and

$$(35.04)\qquad | f(b+iy) - P_2(y) | < \epsilon \qquad (-\infty < y < \infty).$$

If (34.03) holds uniformly over (a, b), and we may apply (34.16), we obtain

$$\begin{aligned}(35.05)\qquad &\frac{\epsilon}{2\pi}\left[\int_{-\infty}^{\infty} | K_1(x-b+i\lambda) | \, d\lambda + \int_{-\infty}^{\infty} | K_2(x-a+i\lambda) | \, d\lambda\right] \\ &\geqq \left| f(x+iy) - \frac{1}{2\pi}\int_{-\infty}^{\infty} P_2(\eta)\, K_1(x+iy-b-i\eta)\, d\eta \right. \\ &\qquad \left. + \frac{1}{2\pi}\int_{-\infty}^{\infty} P_1(\eta)\, K_2(x+iy-a-i\eta)\, d\eta \right| \\ &= \left| f(x+iy) - \frac{1}{2\pi}\sum_1^n B_k \int_{-\infty}^{\infty} e^{iM_k\eta} K_1(x+iy-b-i\eta)\, d\eta \right. \\ &\qquad \left. + \frac{1}{2\pi}\sum_1^n A_k \int_{-\infty}^{\infty} e^{i\Lambda_k\eta} K_2(x+iy-a-i\eta)\, d\eta \right|\end{aligned}$$

$$= \left| f(x+iy) - \frac{1}{2\pi} \sum_1^n B_k e^{iM_k y} \int_{-\infty}^{\infty} e^{iM_k \eta} K_1(x-b-i\eta)\, d\eta \right.$$

$$\left. + \frac{1}{2\pi} \sum_1^n A_k e^{i\Lambda_k y} \int_{-\infty}^{\infty} e^{i\Lambda_k \eta} K_2(x-a-i\eta)\, d\eta \right|$$

$$(35.05) \qquad = \left| f(x+iy) - \sum_1^n B_k e^{M_k(x-b+iy)} \phi_n(M_k) \right.$$

$$\left. + \sum_1^n e^{\Lambda_k(x-a+iy)} (\phi_n(\Lambda_k) - 1) \right|$$

$$= \left| f(x+iy) - \sum_1^{2n} c_k e^{N_k(x+iy)} \right|,$$

where

$$(35.06) \qquad N_{2k-1} = M_k, \qquad N_{2k} = \Lambda_k, \qquad C_{2k-1} = B_k e^{-M_k b} \phi_n(M_k),$$
$$C_{2k} = -A_k e^{-M_k a}(\phi_n(\Lambda_k) - 1).$$

Thus $f(x+iy)$ belongs uniformly to the class of almost periodic functions over $a + \epsilon \leqq x \leqq b - \epsilon$ since,

$$\int_{-\infty}^{\infty} |K_1(x-b+i\lambda)|\, d\lambda = \int_{-\infty}^{\infty} d\lambda \left| \int_{-\infty}^{\infty} \phi_n(y) e^{(x-b+i\lambda)y} dy \right|$$

$$= \int_{-\infty}^{\infty} d\lambda \left| \int_n^{n+2} \frac{x-n}{2} e^{(x-b+i\lambda)y} dy + \int_{n+2}^{\infty} e^{(x-b+i\lambda)y} dy \right|$$

$$(35.07) \qquad = \int_{-\infty}^{\infty} \frac{d\lambda}{2|x-b+i\lambda|} \left| \int_n^{n+2} e^{(x-b+i\lambda)y} dy \right|$$

$$= \int_{-\infty}^{\infty} \frac{d\lambda}{2((x-b)^2+\lambda^2)} \left| e^{(n+2)(x-b+i\lambda)} - e^{n(x-b+i\lambda)} \right|$$

$$\leqq \frac{\text{const.}}{x-b} \leqq \text{const.},$$

$$(35.08) \qquad \int_{-\infty}^{\infty} |K_2(x-a+i\lambda)|\, d\lambda \leqq \text{const.}$$

Chapter IX

Random functions

36. **Random functions.** By a random function we understand a function which, in a certain way to be specified in what follows, depends upon a formally expressed variable and a usually suppressed parameter of distribution. Such functions occur throughout statistical mechanics. In statistical mechanics a certain possible state of the universe is expressed by a function of one or more variables which give the geometrical coordinates and the time, together with a parameter which singles out the universe considered as one among all possible universes and determines its probability. As a more specific example let us take the path of a particle subject to a Brownian motion and let us consider one coordinate of the particle (let us say the x-coordinate) as a function of the time t. Then for any particular Brownian motion or motion of a particle impelled by the collision of neighboring molecules subject to a thermal agitation, x will be a well-defined function of t. If, however, instead of considering the actual course traced by a specified particle, we consider all possible courses traced by all possible particles, in addition to the variable t, x will have an argument which singles out the specific Brownian motion in question from all possible Brownian motions. This variable is introduced for purposes of integration; that is, a certain range of this variable measures by its length the probability of the set of Brownian motions it represents, and an integration with respect to this variable yields a probability average of the quantity integrated. Naturally this probability theory is not at all simple and needs to be specified in detail, and a large part of the remainder of this chapter is devoted to the specification in question.

It will be seen that the theory of random functions is in essentials a theory of integration in function space. As such it may be subsumed under the general theories of integration of Radon, Daniell, and Wiener. A previous attempt at a theory of integration in function space was due to Gâteaux, but this earlier theory is not a special case of the Daniell-Radon integral. The earlier theory attempted to treat every value $F(t_0)$ of a function $F(t)$ as an independent variable. In such a scheme we find that a succession of regions of positive measure may include one another successively in such a way that there is no element common to all of them. This violates one of the most essential canons of the Daniell theory. In order to eliminate it, it is necessary that the class of functions over which we are integrating should be in some sense or other compact; that is, that any sequence of functions should contain a subsequence with a limit function. If this is not strictly true with the original class with which we are dealing, it must be at any rate possible to make it true by removing or adjoining a set of functions of arbitrarily small measure. Compactness in the ordinary sense is equivalent to uniform boundedness and equi-continuity. Now, it is a priori obvious that Gâteaux's type of integration must yield us functions which almost never are continuous, and completely breaks down in this point.

The Brownian motion points us a way out of the difficulty. In this motion it is not the position of a particle which is independent of the position at another time, but the travel of a particle between two stated times which is independent of the position at the first time. In crude language, the differentials of the function $x(t)$ correspond to the independent coordinates of a point in space of a finite number of dimensions, and not the values of the function. Physically it is at least reasonable that the Brownian motion of a particle is continuous, and we shall in fact show that the Einstein theory of the Brownian motion permits us to say that in fact it is almost always continuous. We shall show even more, that it is subject to a condition of equi-continuity except for a set of cases whose measure or what is equivalent, whose probability, we can reduce to as small a value as we want.

The theory of random functions always makes the impression of a much greater degree of artificiality than corresponds to the facts. The reason is that in the theory of Daniell or Lebesgue integration the sum of a denumerable set of null sets is itself null, while the sum of a set of null sets of the power of the continuum need not be null. This fact makes it extremely desirable to characterize a random function by a denumerable set of conditions, while the characterization of such a function by its value for all its arguments yields a number of characterizations of the power of the continuum. If then we attempt to characterize a random function by a set of its values, it is almost necessary for the purposes of our technical development that the original set of values which we take should be denumerable. For example, we may define such a function by its values at the rational or the binary points of the line. On the other hand, we may give up entirely the characterization of such a function by its values and define it by its Fourier coefficients or its coefficients in some other scheme of development. In any case it will turn out that the function which we have so defined will, except in a set of cases of probability zero (or measure zero, which is the same thing), either be continuous or in some readily specified sense be equivalent to a continuous function. Once this is done, we may define the value of this continuous function uniquely over the whole of its interval of definition. We thus have generalized our initial function defined by a denumerable set of parameters to a function apparently defined by a larger number of parameters. In doing so we have made use of a method which masks the true invariance of the result we have obtained. It is, however, a perfectly legitimate mathematical method to obtain our result in a non-invariantive way and to establish its proper invariance later by specific theorems. This mode of procedure however has the disadvantage of being most non-heuristic and of demanding from the reader the patience to take on faith the necessity of a large amount of material whose justification is only given after the completion of the argument. It will ultimately turn out that the random functions we deal with and the measure that properly belongs to them determine a function-space distinct from that of

Hilbert but covariant with it under all unitary transformations which leave Hilbert space invariant. While such a theory may be built up for real Hilbert space, and while this is the case in Wiener's previous writings, the present theory of differential space applies to one covariant with complex Hilbert space.

It will be seen that although the last two chapters of this book are apparently devoted to integration in a continuous infinity of dimensions, they really form chapters in what E. Borel has designated the theory of "denumerable probabilities."* The first close approach to the specific problems which concern us here was made by Steinhaus.† He considers the series

$$\sum_{1}^{\infty} \pm \frac{1}{n}, \tag{36.01}$$

where the signs $\pm$ represent independent choices, and shows that this series converges in almost every case. The methods of Steinhaus were made into a powerful analytic tool by Paley and Zygmund,‡ more especially in the formation of Gegenbeispiele, and the final theory has been employed with great affect by Bohr and Jessen,§ in the study of the Riemann zeta function and of almost periodic functions in the complex domain.

Wiener‖ has developed a theory of random functions in many respects parallel to those discussed by the authors already mentioned, but not identical with them. Since the Wiener theory of random functions is derived from considerations in the theory of the Brownian motion and of statistical mechanics, in which Gaussian distributions play an important rôle, they also play an important rôle in his theory. In this respect, Wiener's theory resembles the Einstein theory of the Brownian motion, to which we shall later show it to be equivalent.¶ Indeed, a sharp distinction may be made between Wiener's theory and all others so far mentioned in that, while in both cases a denumerable set of terms are assigned coefficients with a certain distribution, in all other theories these coefficients either have at random the values ± 1, or are distributed at random around the unit circle in the complex plane. In addition to its applicability to statistical mechanics, Wiener's theory possesses a larger degree of symmetry than the

* E. Borel, *Les probabilités dénombrables et leurs applications arithmétiques*, Rendiconti del Circolo Matematico di Palermo, vol. 27 (1909), pp. 247–271.

† Cf. *Sur la probabilité de la convergence de séries*, Studia Mathematica, vol. 2 (1930), pp. 21–39, and earlier papers there referred to.

‡ *On some series of functions*, Proceedings of the Cambridge Philosophical Society, vol. 26 (1930), pp. 337–357, 458–474; vol. 28, pp. 190–205. Cf. also *Notes on random functions*, Paley, Wiener, and Zygmund, Mathematische Zeitschrift, vol. 37 (1933), pp. 647–688, on which the present chapter is largely based.

§ H. Bohr and B. Jessen, *Über die Werteverteilung der Riemannschen Zetafunktion*, Acta Mathematica, vol. 54 (1930), pp. 1–35; vol. 58 (1932), pp. 1–55.

‖ Cf. N. Wiener, *Generalized harmonic analysis*, Acta Mathematica, vol. 55, pp. 214 ff., and the references there given.

¶ A. Einstein, Annalen der Physik, vol. 17 (1905), pp. 549 ff.; vol. 19 (1906), pp. 371 ff.

alternative theories, or what is the same thing in other words, it has a more extensive group of transformations under which it is invariant. The reason for this is that if a number of terms have independent Gaussian distributions, any linear combination of these terms has itself a Gaussian distribution. On the other hand, if we start with a distribution of the original terms over the values ± 1, or around the unit circle, the distribution of a linear combination of the original terms becomes extremely complex and unmanageable. This property of Gaussian distributions is thus intimately allied with the covariance which the theory of random functions developed here shows with Hilbert space.

By a change of scale, a real Gaussian distribution may be reduced to the distribution of a real parameter uniformly over the interval $(0, 1)$. In the same way, a complex Gaussian distribution may be reduced to the simultaneous and independent uniform distribution of two parameters over an interval of that sort. By this artifice our integration in function space may be carried back to the integration of a function of a denumerable infinity of variables over a cube in its proper space. Such an integration has been discussed by Daniell, Jessen and others.* Since however it is probably not familiar to the reader of this chapter, we shall derive it from the ordinary integration of a function of one variable over an interval of a line by a process of mapping. This process of mapping is merely an elaboration of the process by which a square may be mapped on a line in such a way that planar measures are mapped into equal linear measures.

This mapping of a square on a line with preservation of measure deserves some consideration. Let the coordinates of a point on the unit square $0 < x < 1$, $0 < y < 1$ be represented as binary fractions, in the form

$$(36.02)\qquad \begin{aligned} x &= \alpha_1/2 + \alpha_2/4 + \cdots + \alpha_n/2^n + \cdots \\ &\qquad\qquad (\alpha_n = 0 \text{ or } 1 \text{ for all } n \text{ independently})\,; \\ y &= \beta_1/2 + \beta_2/4 + \cdots + \beta_n/2^n + \cdots \\ &\qquad\qquad (\beta_n = 0 \text{ or } 1 \text{ for all } n \text{ independently})\,. \end{aligned}$$

The correspondence between x and the sequence $\{\alpha_n\}$ is not one-one, but becomes one-one if we disregard all rational values of x with denominators that are powers of two, which have alternative representations terminating in $\cdots 000 \cdots$ and $\cdots 111 \cdots$, respectively. These values of x are a denumerable set, and hence a set of zero linear measure. The points (x, y) with such values of x constitute a set of zero plane measure. Similar considerations apply to the correspondence between y and the sequence $\{\beta_n\}$. Thus except for a set of points (x, y) with

* P. J. Daniell, *Integrals in an infinite number of dimensions*, Annals of Mathematics, (2), vol. 20 (1919), pp. 281–288; B. Jessen, memoir to appear in Acta Mathematica; also, *Bitrag til Integralteorien for Funktioner af unendelig mange Variable*, Copenhagen, 1930.

zero plane measure, the correspondence between the points of the square $0 < x < 1, 0 < y < 1$ and the pair of sequences $\{\alpha_n\}$ and $\{\beta_n\}$ is one-one.

Now, the pair of sequences $\{\alpha_n\}$ and $\{\beta_n\}$ may be rearranged into a single sequence in a large number of different one-one methods. For example, we may rearrange $\{\alpha_n\}$ and $\{\beta_n\}$ in the form

$$\cdot\, \alpha_1 \beta_1 \alpha_2 \beta_2 \alpha_3 \beta_3 \cdots \alpha_n \beta_n \cdots \tag{36.03}$$

and we may assign to it the binary number z with the same representation. The correspondence between z and the sequence (36.03) is one-one except for the null set of rational values of z with denominators which are powers of 2. Furthermore, it is clear that if we determine a sub-class of z in which a denumerable sub-set of digits have fixed values, this set will have zero measure, and the sum of any denumerable set of such sets will also have zero measure. It thus follows that the set of values of z corresponding to pairs (x, y) which do not both have unique binary representations is itself of zero measure. The same is true of the planar measure of the points (x, y) corresponding to values of z for which the binary representation is not unique. Thus, if we discard a planar null set of values of (x, y) and a linear null set of values of z, the mapping of the square $0 < x < 1, 0 < y < 1$ on the interval $0 < z < 1$ determined by (36.03), or any other such rearrangement, is one-one.

The planar set of points determined by

$$\begin{aligned} &\gamma_1/2 + \gamma_2/4 + \cdots + \gamma_m/2^m < x < \gamma_1'/2 + \gamma_2'/4 + \cdots + \gamma_m'/2^m, \\ &\delta_1/2 + \delta_2/4 + \cdots + \delta_n/2^n < y < \delta_1'/2 + \delta_2'/4 + \cdots + \delta_n'/2^n \quad (m > n), \end{aligned} \tag{36.031}$$

where all γ_k's, γ_k''s, δ_k's, and δ_k''s are 0 or 1, consists of $2^{2m}(\cdot\gamma_1' \cdots \gamma_m' - \cdot\gamma_1 \cdots \gamma_m) \times (\cdot\, \delta_1' \cdots \delta_n' - \cdot\, \delta_1 \cdots \delta_n)$ squares, each with an area 2^{-2m}. Transformation (36.03) maps it into the same number of intervals, each of length 2^{-2m}, and with the same total measure. Similarly, any interval of z lying between terminating binaries is the map of a finite set of squares in the (x, y) plane, with total area equal to the length of the interval in question. Thus the transformation (36.02) or any like transformation is measure-preserving, as far as rectangles in the (x, y) plane with terminating binary coordinates for their vertices and intervals in the z line with terminating binary end points are concerned.

Every interval on a line may be included in an interval of arbitrarily slightly greater measure, but with terminating binary end points, and includes such an interval of arbitrarily slightly less measure. In the plane, we have an exactly similar theorem concerning intervals. It thus follows at once that the mapping of (36.03) or any similar mapping transforms every rectangle in the plane into a measurable set on the line with linear measure equal to the area of the rectangle, and that every interval on the line is the transform of a measurable set in the plane with a planar measure equal to the linear measure of the interval. By methods familiar in the theory of the Lebesgue integral, it follows that to every measurable set in the plane unit rectangle, there corresponds a measurable set on the unit interval with linear measure equal to the planar measure of the original set, and that every measurable set on the unit interval corresponds to a planar measurable set in the unit rectangle with a plane measure equal to the linear measure of the original set. The whole theory of Lebesgue integration in two dimensions may thus be derived by this mapping from the corresponding theory in one dimension. Given any function over the unit rectangle, our process of mapping determines an image function over the unit interval, and if this is integrable, its integral will agree with the integral of the original function of two variables, *and might have been used to define that integral.* In an exactly similar manner, regions in three or more dimensions may be mapped on a segment of a line with preservation of measure, and this mapping might have been used to establish the definition of the Lebesgue integral in n dimensions. This method of definition has the distinct advantage, that all theorems concerning Lebesgue integration in one dimension may be transferred directly, without any modification of their proofs, to Lebesgue integration in n dimensions.

This consideration is even more important when it comes to the establishment of the Lebesgue integral in a space of a denumerable infinity of dimensions. As we have seen, Daniell has established a general theory of integration, and he has subsumed* under this as a special case integration in a denumerable infinity of dimensions. The advantage of this method is logical straightforwardness and transparency; the disadvantage, that we cannot call upon the great body of existing theorems in the theory of the Lebesgue integral, but must establish each one de novo by the trivial modification of its original method of proof. In an ideal course on Lebesgue integration, all theorems would be developed from the point of view of the Daniell integral; but in the present state of mathematical education, a method of mapping has distinct advantages.

The mapping of the square on the linear segment has given a sufficient exposition of the general principles of mapping for us to dispense with a discussion of the biunivocal character of the mapping we are now developing. This mapping has as its purpose the representation on the segment $0 < \alpha < 1$ the points $(\alpha_1, \alpha_2, \cdots, \alpha_n, \cdots)$ of the region

$$\left\{\begin{array}{l} 0 < \alpha_1 < 1; \\ 0 < \alpha_2 < 1; \\ \cdots\cdots\cdots\cdots \\ 0 < \alpha_n < 1; \\ \cdots\cdots\cdots\cdots \end{array}\right. \tag{36.032}$$

of space of a denumerable infinity of dimensions. Let the binary representation of α_k be

$$\cdot\, \alpha_{k1}\, \alpha_{k2} \cdots \alpha_{kn} \cdots, \tag{36.04}$$

* P. J. Daniell, *A general form of integral*, Annals of Mathematics, (2), vol. 19 (1918), pp. 279–294.

and let us put

(36.05) $$\alpha = \cdot\, \alpha_{11}\,\alpha_{12}\,\alpha_{21}\,\alpha_{13}\,\alpha_{22}\,\alpha_{31}\,\alpha_{14}\,\alpha_{23}\,\alpha_{32}\,\alpha_{41}\cdots.$$

Let

(36.06) $$F(\alpha) = f(\alpha_1, \alpha_2, \cdots, \alpha_\nu)$$

be an integrable function of a finite number of the α_k's. Then by methods substantially the same as those we have used in the discussion of the mapping of the square on the segment, we have

(36.07) $$\int_0^1 F(\alpha)\,d\alpha = \int_0^1 d\alpha_1 \cdots \int_0^1 d\alpha_\nu\, f(\alpha_1, \cdots, \alpha_\nu).$$

This follows at once from the well known fact concerning the Lebesgue integral of a function of a single variable, that if $f(x)$ is a summable function of x over the range (0, 1), then if $\epsilon > 0$, there exists a step-function $g(x)$ such that

(36.08) $$\int_0^1 |f(x) - g(x)|\,dx < \epsilon.$$

There is no difficulty whatever in showing that the function g may be so chosen that the abscissae of all the jumps are terminating binaries. If we apply this result to functions of $(\alpha_1, \alpha_2, \cdots)$, and if we identify integration over $(\alpha_1, \alpha_2, \cdots)$ with integration of the corresponding function over α, we see that every function $f(\alpha_1, \alpha_2, \cdots, \alpha_n, \cdots)$ for which

$$\int_0^1 d\alpha_1 \int_0^1 d\alpha_2 \cdots \int_0^1 d\alpha_n \cdots f(\alpha_1, \alpha_2, \cdots, \alpha_n, \cdots)$$

exists determines at least one step-function $g(\alpha_1, \alpha_2, \cdots, \alpha_\nu)$ of the finite set of variables $\alpha_1, \cdots, \alpha_\nu$, such that

(36.09) $$\int_0^1 d\alpha_1 \int_0^1 d\alpha_2 \cdots \int_0^1 d\alpha_n \cdots |f(\alpha_1, \alpha_2, \cdots, \alpha_n, \cdots) - g(\alpha_1, \alpha_2, \cdots, \alpha_\nu)| < \epsilon.$$

37. **The fundamental random function.** Up to the present, we have been discussing an infinity of distinct real variables α_ν, uniformly and independently distributed over the interval (0, 1). We wish to transfer this discussion to an infinity of complex variables, the real and the imaginary parts of which have independent Gaussian distributions. To this end we consider $(-\log \alpha_j)^{1/2} e^{2\pi i \alpha_k}$, where $0 \leqq \alpha_j \leqq 1$, $0 \leqq \alpha_k \leqq 1$, $j \neq k$. If we put

(37.01) $$r = (-\log \alpha_j)^{1/2}, \qquad \theta = 2\pi\,\alpha_k,$$

we have

(37.02) $$|d\alpha_j\,d\alpha_k| = \left|\frac{1}{\pi} re^{-r^2}\,dr\,d\theta\right| \qquad (0 \leqq r < \infty, \quad 0 \leqq \theta \leqq 2\pi).$$

If now

$$x = r\cos\theta\,, \qquad y = r\sin\theta\,, \tag{37.03}$$

we have

$$|\,d\alpha_j\, d\alpha_k\,| = \frac{1}{\pi}\, e^{-(x^2+y^2)}\,|\,dx\,dy\,| \qquad (-\infty < x < \infty, \quad -\infty < y < \infty)\,. \tag{37.04}$$

Thus if α_j and α_k are uniformly distributed over (0, 1) and are independent, the real and the imaginary parts of $(-\log\alpha_j)^{1/2}\, e^{2\pi i \alpha_k}$ have independent Gaussian distributions.

Let us now consider the formal trigonometric series

$$\sum_{-\infty}^{\infty} a_k\, e^{ikx} \tag{37.05}$$

where the real and the imaginary parts of each a_k all have the same Gaussian distribution, and are all independent of one another. This series will almost never converge, even in the mean, but we shall show that its formal integral,

$$\psi(x, \alpha) \sim x a_0 + \sum_1^{\infty} \frac{a_n\, e^{inx}}{in} + \sum_1^{\infty} \frac{a_{-n}\, e^{-inx}}{-in}\,, \tag{37.06}$$

will exist as a limit in the mean for almost all choices of $\{a_k\}$. We shall express this more precisely by saying that

$$\begin{aligned} \psi(x,\alpha) \sim x(-\log\alpha_1)^{1/2} e^{2\pi i\alpha_2} &+ \sum_1^{\infty} \frac{e^{inx}}{in}(-\log\alpha_{4n-1})^{1/2} e^{2\pi i \alpha_{4n}} \\ &+ \sum_1^{\infty} \frac{e^{-inx}}{-in}(-\log\alpha_{4n+1})^{1/2} e^{2\pi i\alpha_{4n+2}} \end{aligned} \tag{37.061}$$

exists as a limit in the mean with respect to x, for almost all α. To this end, we must show that for almost all α,

$$-\sum_1^{\infty} \frac{1}{n^2}\log\alpha_{4n-1} - \sum_1^{\infty}\frac{1}{n^2}\log\alpha_{4n+1} < \infty\,. \tag{37.062}$$

We shall first investigate the distribution of

$$-\sum_M^N \frac{1}{n^2}\log\alpha_{4n-1} - \sum_M^N \frac{1}{n^2}\log\alpha_{4n+1}\,. \tag{37.063}$$

By (36.07)

$$\begin{aligned} \int_0^1 \left(-\sum_M^N \frac{1}{n^2}\log\alpha_{4n-1} - \sum_M^N\frac{1}{n^2}\log\alpha_{4n+1}\right) d\alpha \\ = 2\sum_M^N \frac{1}{n^2} = O\left(\frac{1}{M}\right) \qquad (N > M)\,. \end{aligned} \tag{37.07}$$

Again,

$$\int_0^1 \left[\sum_M^N \frac{1}{n^2}\log \alpha_{4n-1} + \sum_M^N \frac{1}{n^2}\log \alpha_{4n+1} + 2\sum_M^N \frac{1}{n^2}\right]^2 d\alpha \tag{37.08}$$
$$= 6\sum_M^N \frac{1}{n^4} = O\left(\frac{1}{M^3}\right).$$

From this it follows that except over a set of values of α of measure not exceeding some C_2/M,

$$-\sum_M^N \frac{1}{n^2}\log \alpha_{4n-1} - \sum_M^N \frac{1}{n^2}\log \alpha_{4n+1} < C_1/M. \tag{37.09}$$

If we now let $M = \nu^2$, $N = (\nu + 1)^2$, we see that for almost all values of α, there exists a C_1, depending in general upon α, such that

$$-\sum_1^\infty \frac{1}{n^2}\log \alpha_{4n-1} - \sum_1^\infty \frac{1}{n^2}\log \alpha_{4n+1} \leqq \sum_1^\infty C_1/\nu^2 < \infty. \tag{37.10}$$

Thus $\psi(x, \alpha)$ is determined for almost all x, for almost all α, and belongs to L_2 for almost all α.

We wish, however, to prove the stronger

THEOREM XLIII. *A sequence of partial sums of the right-hand side of* (37.061) *converges uniformly in x to a limit for almost all α. Thus $\psi(x, \alpha)$ may so be defined as to be continuous in x for almost all α:*

$$\psi_{m,n}(x, \alpha) = \sum_{m+1}^n \frac{e^{ikx}}{ik}(-\log \alpha_{4k-1})^{1/2} e^{2\pi i \alpha_{4k}}. \tag{37.11}$$

We shall prove that

$$\sum_1^\infty |\psi_{2^n, 2^{n+1}}(x, \alpha)| \tag{37.12}$$

converges uniformly for almost all α, and hence that

$$\sum_1^\infty \psi_{2^n, 2^{n+1}}(x, \alpha) \tag{37.13}$$

converges uniformly for almost all α. It will follow by the same reasoning that

$$\sum_1^\infty \left[\sum_{2^n+1}^{2^{n+1}} \frac{e^{-ikx}}{ik}(-\log \alpha_{4k-3})^{1/2} e^{2\pi i \alpha_{4k+2}}\right] \tag{37.14}$$

converges uniformly for almost all α, and Theorem XLIII will be established.

We have

$$
\begin{aligned}
|\psi_{m,n}(x,\alpha)|^2 &= \sum_{m+1}^{n} \frac{e^{ikx}}{k}(-\log \alpha_{4k-1})^{1/2} e^{2\pi i \alpha_{4k}} \sum_{m+1}^{n} \frac{e^{-ilx}}{l}(-\log \alpha_{4l-1})^{1/2} e^{-2\pi i \alpha_{4l}} \\
&= \sum_{m+1}^{n} \frac{-\log_{4k-1}}{k^2} \\
&\quad + 2\Re\left\{ \sum_{j=1}^{n-m-1} e^{ijx} \sum_{k=m+1+j}^{n} (-\log \alpha_{4k-1})^{1/2}(-\log \alpha_{4(k-j)-1})^{1/2} \right. \\
&\qquad\qquad \left. \times [\exp(2\pi i(\alpha_{4k} - \alpha_{4(k-j)}))] \right\} \\
&\leqq \frac{1}{m^2}\sum_{m+1}^{n}(-\log \alpha_{4k-1}) + 2\sum_{j=1}^{n-m-1}\left| \sum_{k=m+1+j}^{n} (-\log \alpha_{4k-1})^{1/2} \right. \\
&\qquad\qquad \left. \times (-\log \alpha_{4(k-j)-1})^{1/2} \exp(2\pi i(\alpha_{4k} - \alpha_{4(k-j)})) \right| .
\end{aligned}
\tag{37.15}
$$

Thus

$$
\begin{aligned}
&\int_0^1 \text{l.u.b.}\, |\psi_{m,n}(x,\alpha)|^2 \, d\alpha \\
&\leqq \frac{n}{m^2}\int_0^1 (-\log \xi)\, d\xi \\
&\quad + 2\sum_{j=1}^{n-m-1} \int_0^1 d\xi_1 \cdots \int_0^1 d\xi_{n-m-1} \int_0^1 d\eta_1 \cdots \int_0^1 d\eta_{n-m-1} \\
&\qquad \times \left| \sum_{k=m+1+j}^{n} (\log \xi_k \log \xi_{k-j})^{1/2} \exp(2\pi i\, \eta_k) \right| \\
&\leqq \frac{n}{m^2} + n^2 m^\epsilon \exp(-m^\epsilon) \\
&\quad + 2\sum_{j=1}^{n-m-1} \int_{\exp(-m^\epsilon)}^1 d\xi_1 \cdots \int_{\exp(-m^\epsilon)}^1 d\xi_{n-m-1} \int_0^1 d\eta_1 \cdots \int_0^1 d\eta_{n-m-1} \\
&\qquad \times \left| \sum_{k=m+1+j}^{n} \frac{(\log \xi_k \log \xi_{k-j})^{1/2} \exp(2\pi i\, \eta_k)}{k(k-j)} \right| ,
\end{aligned}
\tag{37.16}
$$

where ϵ is some positive quantity to be fixed later.

Now, by the Schwarz inequality,

$$
(37.17)\quad
\begin{aligned}
\int_0^1 d\eta_1 \cdots \int_0^1 d\eta_p \left| \sum_1^p a_k\, e^{2\pi i \eta_k} \right|
&\leqq \left\{ \int_0^1 d\eta_1 \cdots \int_0^1 d\eta_p \left| \sum_1^p a_k\, e^{2\pi i \eta_k} \right|^2 \right\}^{1/2} \\
&= \left\{ \int_0^1 d\eta_1 \cdots \int_0^1 d\eta_p \sum_{j=1}^p \sum_{k=1}^p a_j\, \bar{a}_k\, e^{2\pi i (\eta_j - \eta_k)} \right\}^{1/2} \\
&= \left\{ \sum_{k=1}^p |\, a_k \,|^2 \right\}^{1/2}.
\end{aligned}
$$

Applying this to (37.16) we get

$$
(37.18)\quad
\begin{aligned}
\int_0^1 \underset{x}{\text{l.u.b.}}\, |\psi_{m,2m}(x,\alpha)|^2\, d\alpha &\leqq O\left(\frac{1}{m}\right) + 2 \sum_{j=1}^{m-1} \left[\sum_{k=m+1+j}^{2m} \frac{m^{2\epsilon}}{[k(k-j)]^2} \right]^{1/2} \\
&= O\left(\frac{1}{m}\right) + 2 \sum_{j=1}^{m-1} \left[O\left(\int_{m+j}^{2m} \frac{m^{2\epsilon}}{x^2\,(x-j)^2}\, dx \right) \right]^{1/2} \\
&\leqq O\left(\frac{1}{m}\right) + 2 \sum_{j=1}^{m-1} \left[O\left(\int_m^{3m} \frac{m^{2\epsilon}}{x^4}\, dx \right) \right]^{1/2} \\
&= O\left(\frac{1}{m}\right) + O\left(\frac{m}{m^{3/2-\epsilon}}\right) = O\,(m^{-1/4})
\end{aligned}
$$

if $\epsilon < \frac{1}{4}$. Thus except over a set of values of α of measure not exceeding $cm^{-1/12}$ we have

$$
(37.19)\qquad \underset{x}{\text{l.u.b.}}\, |\, \psi_{m,\,2m}(x,\alpha) \,| \leqq c\, m^{-1/12},
$$

and in particular, except over a set of values of α of measure not exceeding $c_1\, 2^{-n/12}$ we have

$$
(37.20)\qquad \underset{x}{\text{l.u.b.}}\, |\, \psi_{2^n,\,2^{n+1}}(x,\alpha) \,| < c_1\, 2^{-n/12}.
$$

It follows that except over a set of values of α of measure not exceeding

$$
(37.21)\qquad c_1 \sum_N^\infty 2^{-n/12} = \frac{c_1\, 2^{-N/12}}{1 - 2^{-1/12}}
$$

we have, for all x,

$$
(37.22)\qquad \left| \sum_N^\infty \psi_{2^n,\,2^{n+1}}(x,\alpha) \right| \leqq c_1 \sum_N^\infty 2^{-n/12} = \frac{c_1\, 2^{-N/12}}{1 - 2^{-1/12}};
$$

and since the expression of (37.21) tends to 0 as $N \to \infty$ we see that except over a null set of values of α,

$$\lim_{N\to\infty} \sum_{1}^{N} \psi_{2^n, 2^{n+1}}(x, \alpha) \tag{37.23}$$

exists uniformly in x. This establishes Theorem XLIII. Here we use the familiar fact that if a series converges to one limit and converges in the mean to another, these limits differ at most over a null set.

In all that follows, we take $\psi(x, \alpha)$ to be continuous. Now let $F(x)$ belong to L_2, and let

$$F(x) = \sum_{-n}^{n} f_k e^{ikx}. \tag{37.24}$$

We wish to define $\int_{-\pi}^{\pi} F(x)\, d\psi(x, \alpha)$ but cannot define it as an ordinary Stieltjes integral, as we have no reason to believe that $\psi(x, \alpha)$ is almost always of limited total variation in x; and as a matter of fact it is almost never of limited total variation. However, let us define the integral by carrying out the indicated integration by parts. We have

$$\begin{aligned}\int_{-\pi}^{\pi} F(x)\, d\psi(x, \alpha) &= F(\pi)\, \psi(\pi, \alpha) - F(-\pi)\, \psi(-\pi, \alpha) \\ &\qquad - \int_{-\pi}^{\pi} \psi(x, \alpha)\, F'(x)\, dx \\ &= 2\pi \Big\{ (-\log \alpha_1)^{1/2} e^{2\pi i \alpha_2} f_0 + \sum_{1}^{N} (-\log \alpha_{4k-1})^{1/2} e^{2\pi i \alpha_{4k}} f_k \\ &\qquad + \sum_{1}^{N} (-\log \alpha_{4k+1})^{1/2} e^{2\pi i \alpha_{4k+2}} f_{-k} \Big\}.\end{aligned} \tag{37.25}$$

We now wish to assert and prove

THEOREM XLIV. *Let $F(x)$ be defined as in (37.24). Let Φ be any function for which*

$$\frac{1}{2\pi} \int_{-\pi}^{\pi} d\theta \int_{0}^{\infty} e^{-u}\, \Phi \left\{ e^{i\theta} \left[2\pi u \int_{-\pi}^{\pi} |F(x)|^2\, dx \right]^{1/2} \right\} du = I \tag{37.26}$$

exists as an absolutely convergent Lebesgue integral. Then

$$\int_{0}^{1} \Phi \left\{ \int_{-\pi}^{\pi} F(x)\, d\psi(x, \alpha) \right\} d\alpha = I. \tag{37.27}$$

Similarly, let

$$G(x) = \sum_{-n}^{n} g_k e^{ikx}. \tag{37.28}$$

Let

$$\int_{-\pi}^{\pi} |F(x)|^2 dx = A, \quad \int_{-\pi}^{\pi} \bar{F}(x) G(x) dx = B, \quad \int_{-\pi}^{\pi} |G(x)|^2 dx = C, \tag{37.29}$$
$$AC - |B|^2 = D,$$

and let Φ *be any function for which*

$$\begin{aligned} &\frac{1}{4\pi^4 D} \int_{-\infty}^{\infty} du_1 \int_{-\infty}^{\infty} du_2 \int_{-\infty}^{\infty} dv_1 \int_{-\infty}^{\infty} dv_2 \\ &\qquad \times \exp\left[\frac{-(u_1^2+v_1^2)C + 2(u_1u_2+v_1v_2)\Re(B) - (u_2^2+v_2^2)A + 2(u_1v_2-u_2v_1)\Im(B)}{2\pi D}\right] \\ &\qquad \times \Phi(u_1+iv_1, u_2+iv_2) = J \end{aligned} \tag{37.30}$$

exists as an absolutely convergent Lebesgue integral. Then

$$\int_0^1 \Phi\left\{\int_{-\pi}^{\pi} F(x)\, d\psi(x,\alpha), \int_{-\pi}^{\pi} G(x)\, d\psi(x,\alpha)\right\} d\alpha = J. \tag{37.301}$$

In particular, if $F(x)$ *and* $G(x)$ *are orthogonal to each other in the complex sense,* $B = 0$ *and* $D = AC$. *Thus*

$$\begin{aligned} J &= \frac{1}{4\pi^4 AC} \int_{-\infty}^{\infty} du_1 \int_{-\infty}^{\infty} du_2 \int_{-\infty}^{\infty} dv_1 \int_{-\infty}^{\infty} dv_2 \\ &\qquad \times \exp\left\{-\frac{u_1^2+v_1^2}{A} - \frac{u_2^2+v_2^2}{C}\right\} \Phi(u_1+iv_1, u_2+iv_2) \\ &= \int_0^1 d\alpha \int_0^1 d\beta\, \Phi\left\{\int_{-\pi}^{\pi} F(x)\, d\psi(x,\alpha), \int_{-\pi}^{\pi} G(x)\, d\psi(x,\beta)\right\}. \end{aligned} \tag{37.302}$$

To prove the equivalence of (37.26) and (37.27), let us notice that

$$\begin{aligned} I &= \frac{1}{\pi} \int_{-\infty}^{\infty} dx \int_{-\infty}^{\infty} dy\, \Phi\left\{2\pi\left[\sum_{-n}^{n} |f_k|^2\right]^{1/2} (x+iy)\right\} e^{-x^2-y^2} \\ &= \frac{1}{\pi^{2n+1}} \int_{-\infty}^{\infty} dx_{-n} \cdots \int_{-\infty}^{\infty} dx_n \int_{-\infty}^{\infty} dy_{-n} \cdots \int_{-\infty}^{\infty} dy_n \\ &\qquad \times \exp\left(-\sum_{-n}^{n} x_k^2 - \sum_{-n}^{n} y_k^2\right) \Phi\left\{2\pi \sum_{-n}^{n} f_k(x_k+iy_k)\right\}. \end{aligned} \tag{37.303}$$

Here we make use of the facts that

$$\frac{1}{\pi^{1/2}} \int_{-\infty}^{\infty} e^{-x^2}\, dx = 1, \tag{37.304}$$

and that

$$\exp\left(-\sum_{-n}^{n} x_k^2 - \sum_{-n}^{n} y_k^2\right)$$

is invariant under a unitary transformation of the variables $x_k + iy_k$. By a simple change in variables,

$$\begin{aligned} I = \int_0^1 d\alpha_1 \cdots \int_0^1 d\alpha_{4n+2}\, \Phi \Bigg\{ 2\pi \Bigg[(-\log \alpha_1)^{1/2} e^{2\pi i \alpha_2} f_0 \\ + \sum_1^n (-\log \alpha_{4k-1})^{1/2} e^{2\pi i \alpha_{4k}} f_k + \sum_1^n (-\log \alpha_{4k+1})^{1/2} e^{2\pi i \alpha_{4k+2}} f_{-k} \Bigg] \Bigg\}, \end{aligned} \tag{37.305}$$

from which (37.27) follows by (37.25).

The proof of (37.301) is quite similar. By the indicated change of variables, we have

$$\begin{aligned} J &= \frac{1}{\pi^2} \int_{-\infty}^{\infty} dx_1 \int_{-\infty}^{\infty} dy_1 \int_{-\infty}^{\infty} dx_2 \int_{-\infty}^{\infty} dy_2 \exp(-x_1^2 - y_1^2 - x_2^2 - y_2^2) \\ &\quad \times \Phi \left\{ (2\pi A)^{1/2} (x_1 + iy_1), \frac{(2\pi)^{1/2}}{A^{1/2}} B(x_1 + iy_1) + \left(2\pi \frac{D}{A}\right)^{1/2} (x_2 + iy_2) \right\} \\ &= \frac{1}{\pi^{2n+1}} \int_{-\infty}^{\infty} dx_{-n} \cdots \int_{-\infty}^{\infty} dx_n \int_{-\infty}^{\infty} dy_{-n} \cdots \int_{-\infty}^{\infty} dy_n \exp\left(-\sum_{-n}^{n} x_k^2 - \sum_{-n}^{n} y_k^2\right) \\ &\quad \times \Phi \left\{ 2\pi \sum_{-n}^{n} f_k (x_k + iy_k),\ 2\pi \sum_{-n}^{n} g_k (x_k + iy_k) \right\}, \end{aligned} \tag{37.306}$$

from which (37.30) follows as (37.27) follows from (37.303).

Formula (37.302) is merely a special case of (37.30), combined with (37.27) on $F(x)$ and $G(x)$ simultaneously. It means that if the two functions $F(x)$ and $G(x)$ are orthogonal, the two expressions

$$\int_{-\pi}^{\pi} F(x)\, d\psi(x, \alpha) \qquad \text{and} \qquad \int_{-\pi}^{\pi} G(x)\, d\psi(x, \alpha)$$

are completely independent, not merely linearly independent. This result is readily extended to any finite set $\{G_n(x)\}$ of orthogonal functions with terminating Fourier developments.

We now wish to eliminate this last restriction that the Fourier developments of F and G or G_n terminate. We do this by the following:

THEOREM XLV. *Let*

$$F(x) \sim \sum_{-\infty}^{\infty} f_k e^{ikx}, \tag{37.31}$$

and let

$$\sum_{-\infty}^{\infty} | f_k |^2 < \infty . \tag{37.32}$$

Let

$$F_n(x) = \sum_{-n}^{n} f_k e^{ikx} . \tag{37.33}$$

Then, except for a set of values of α of zero measure,

$$\lim_{n\to\infty} \int_{-\pi}^{\pi} F_n(x)\, d\psi(x, \alpha) \tag{37.34}$$

exists. We shall define $\int_{-\pi}^{\pi} F(x)\, d\psi(x, \alpha)$ to be this latter limit.

In proving this, let us first note that an argument exactly similar to that which we have used in proving (37.062) shows that for almost all α,

$$- \log \alpha_1 | f_0 |^2 - \sum_{1}^{n} (- \log \alpha_{4k-1}) | f_k |^2 - \sum_{1}^{n} (- \log \alpha_{4k+1}) | f_{-k} |^2 < \infty . \tag{37.35}$$

We now appeal to a result of Rademacher,* to the effect that if

$$\sum_{1}^{\infty} | c_n |^2 < \infty , \tag{37.36}$$

then for almost all choices of the sequence of signs,

$$\left| \sum_{1}^{\infty} \pm c_n \right| < \infty . \tag{37.37}$$

To prove this, we introduce the Rademacher functions

$$\phi_n(x) = (-1)^{[2^n x]} . \tag{37.38}$$

We wish to show that for almost all x, $\sum_1^\infty c_n \phi_n(x)$ converges. To begin with, the functions $\phi_n(x)$ are normal and orthogonal, and there is a function $\phi(x)$ belonging to L_2 such that

$$\lim_{n\to\infty} \int_0^1 \left| \phi(x) - \sum_{1}^{n} c_k \phi_k(x) \right|^2 dx = 0 . \tag{37.39}$$

* H. Rademacher, *Einige Sätze über Reihen von allgemeinen Orthogonalfunktionen*, Mathematische Annalen, vol. 87 (1922), pp. 112–138.

Let us notice that if $2^{-m}[2^m x] = x_m$,

$$2^m \int_{x_m}^{x_m+2^{-m}} \phi(\xi)\, d\xi = \sum_1^m c_n \phi_n(x)\,. \tag{37.40}$$

Thus what we wish to show is that for almost all x,

$$\phi(x) = \lim_{m\to\infty} 2^m \int_{x_m}^{x_m+2^{-m}} \phi(\xi)\, d\xi\,. \tag{37.41}$$

Let it be noted that this is very close to the fundamental theorem of the calculus, to the effect that for almost all ϵ,

$$\phi(x) = \lim_{\epsilon\to 0} \epsilon^{-1} \int_x^{x+\epsilon} \phi(\xi)\, d\xi\,. \tag{37.42}$$

As a matter of fact, (37.41) is an immediate corollary of (37.42), for

$$\begin{aligned} 2^m \int_{x_m}^{x_m+2^{-m}} \phi(\xi)\, d\xi = {} & \frac{1}{x_m + 2^{-m} - x} \int_x^{x_m+2^{-m}} \phi(\xi)\, d\xi\, \frac{x_m + 2^{-m} - x}{2^{-m}} \\ & + \frac{1}{x - x_m} \int_{x_m}^{x} \phi(\xi)\, d\xi\, \frac{x - x_m}{2^{-m}}\,. \end{aligned} \tag{37.43}$$

We now appeal to the fact that the measure of the set of values of x for which any p Rademacher functions have an assigned sign is exactly 2^{-p}, which is equal to the probability that p independent choices between equally probable alternatives yield a certain definite set of results. We may accordingly exchange the measure of a certain set of values of x for which the Rademacher functions have assigned signs with the probability a priori that the sequence of signs be as indicated. This establishes Rademacher's theorem.

If we apply Rademacher's theorem to

$$\begin{aligned} \pm (-\log \alpha_1)^{1/2} f_0 \cos 2\pi\alpha_2 + \sum_1^\infty \pm (-\log \alpha_{4k-1})^{1/2} f_k \cos 2\pi\alpha_{4k} \\ + \sum_1^\infty \pm (-\log \alpha_{4k+1})^{1/2} f_{-k} \cos 2\pi\alpha_{4k+2} \end{aligned} \tag{37.44}$$

and

$$\begin{aligned} \pm (-\log \alpha_1)^{1/2} f_0 \sin 2\pi\alpha_2 + \sum_1^\infty \pm (-\log \alpha_{4k-1})^{1/2} f_k \sin 2\pi\alpha_{4k} \\ + \sum_1^\infty \pm (-\log \alpha_{4k+1})^{1/2} f_{-k} \sin 2\pi\alpha_{4k+2}\,, \end{aligned} \tag{37.45}$$

we see that for almost all values of α, the sequences of the partial sums of (37.44) and (37.45) converge for almost all sequences of signs $\pm$. However,

$$\cos 2\pi(\alpha_\nu + \tfrac{1}{2}) = -\cos 2\pi\alpha_\nu, \qquad \sin 2\pi(\alpha_\nu + \tfrac{1}{2}) = -\sin 2\pi\alpha_\nu. \tag{37.451}$$

Thus a change of signs of the terms of (37.44) or (37.45) may be taken as the replacement of the set of variables $\{\alpha_\nu\}$ which are independently distributed over (0, 1) by another set $\{\beta_\nu\}$ of independent variables with the same distribution. It can accordingly have no effect on the integral or average of any function of the $\{\alpha_\nu\}$. This is not merely true of a *particular* change in the signs of the terms: it is equally true of *almost all* changes of sign of the terms, specified in such a way as to have a well-defined distribution. Hence if we use the definition (37.25), Theorem XLV follows at once.

It immediately follows that

THEOREM XLIV′. *All the results of Theorem* XLIV *may be extended to any functions F and G lying in L_2, whether their Fourier series terminate or not.*

Here we make use of the fact we have already proved, that any functional of infinitely many variables belonging to L_1 may be approximated L_1 by a function of a finite number of these variables.

As a particular application of Theorem XLIV′, let $-\pi \leqq a \leqq b \leqq c \leqq d \leqq \pi$ and let

$$\begin{cases} f_1(x) = 1 \text{ over } (a, b)\,; & f_1(x) = 0 \text{ otherwise}\,; \\ f_2(x) = 1 \text{ over } (c, d)\,; & f_2(x) = 0 \text{ otherwise}\,. \end{cases} \tag{37.46}$$

Then clearly

$$\int_{-\pi}^{\pi} f_1(x) f_2(x)\, dx = 0\,. \tag{37.47}$$

Hence

$$\int_0^1 \Phi\left\{\int_{-\pi}^{\pi} f_1(x)\, d\psi(x, \alpha)\right\} d\alpha = \frac{1}{2\pi}\int_{-\pi}^{\pi} d\theta \int_0^\infty e^{-u}\,\Phi\{[2\pi u e^{2i\theta}(b-a)]^{1/2}\}\, du\,, \tag{37.48}$$

and

$$\begin{aligned} &\int_0^1 \Phi\left\{\int_{-\pi}^{\pi} f_1(x)\, d\psi(x, \alpha),\ \int_{-\pi}^{\pi} f_2(x)\, d\psi(x, \alpha)\right\} d\alpha \\ &= \frac{1}{4\pi^2}\int_{-\pi}^{\pi} d\theta_1 \int_{-\pi}^{\pi} d\theta_2 \int_0^\infty e^{-u_1}\, du_1 \int_0^\infty e^{-u_2}\,\Phi\{[2\pi u_1 e^{2i\theta_1}(b-a)]^{1/2}, \\ &\qquad [2\pi u_2 e^{2i\theta_2}(d-c)]^{1/2}\}\, du_2\,. \end{aligned} \tag{37.49}$$

We obtain similar results for functions of more arguments.

We may state (37.48) and (37.49) as follows: if $y = \psi(x, \alpha)$ is a curve dependent on α as a parameter of distribution, the distribution of $y_2 - y_1$ is dependent only on the corresponding $x_2 - x_1$, and not on the "past" or "future" of x or y.

If t is the time, $\psi(t, \alpha)$ represents one coordinate of a particle subject to a random but uniformly distributed sequence of impulses, such as we find in the Brownian motion, according to the famous theory of Einstein* and Smoluchowski.†

38. **The continuity properties of a random function.** The anomalous properties of the Brownian motion are familiar to the physicist. In the words of Perrin,‡ "This mechanism (of the Brownian movement) has been subject to a detailed analysis by Einstein, in an admirable series of theoretical papers. The approximate but very suggestive analysis given by Smoluchowski certainly deserves to be mentioned also.

"Einstein and Smoluchowski have defined the activity of the Brownian movement in the same way. Previously we had been obliged to determine the 'mean velocity of agitation' by following as nearly as possible the path of a grain. Values so obtained were always a few microns per second for grains of the order of a micron.

"But such evaluations of the activity are *absolutely wrong.* The trajectories are confused and complicated so often and so rapidly that it is impossible to follow them: the trajectory actually measured is very much simpler and shorter than the real one. Similarly, the apparent mean speed of a grain during a given time varies *in the wildest way* in magnitude and direction, and does not tend to a limit as the time taken for observation decreases, as may easily be shown by noting, in the camera lucida, the positions occupied by a grain from minute to minute, and then every five seconds, or, better still, by photographing them every twentieth of a second, as has been done by Victor Henri, Comandon, and de Broglie when kinematographing the movement. It is impossible to fix a tangent, even approximately, and we are thus reminded of the continuous underived functions of the mathematicians. It would be incorrect to regard such functions as mere mathematical curiosities, since indications are to be found in nature of 'underived' as well as 'derived' processes . . . .

"In accordance with the conclusions arrived at from qualitative observation, we shall regard the Brownian movement as *completely irregular* in all directions at right angles to the vertical . . . .

"It must be borne in mind, however, that this result ceases to be exact when the times become so short that the movement is not absolutely irregular. This must necessarily be so, otherwise the *true* velocity would be infinite. *The minimum time within which irregularity may be expected* is probably of the same order as the time . . . . (which) elapses between successive molecular impacts . . . . "

It is most interesting that Perrin finds the Brownian movement reminiscent of non-differentiable continuous functions. In the first instance, he is speaking

* A. Einstein, loc. cit.

† M. von Smoluchowski, Die Naturwissenschaften, vol. 6 (1918), pp. 253–263.

‡ J. Perrin, *Atoms.* Translated by D. Ll. Hammick. Second English edition, London, 1923, pp. 109 ff.

of this movement as a geometrical movement, in which x is plotted against y, but the same question arises when we regard x as well as y, or the complex variable $x + iy$, as a function of the time t. It is a comparatively simple matter to prove that the probability that such a function shall have a derivative for any special fixed argument is zero. We shall prove the much more difficult and general theorem, to the effect that almost all such functions fail to have a derivative for any argument whatever. This is a corollary of the even more general*

THEOREM XLVI. *The values of α for which there exists a t such that*

$$\overline{\lim_{\epsilon \to 0}} \, | \psi(t + \epsilon, \alpha) - \psi(t, \alpha) | \, \epsilon^{-\lambda} < \infty \qquad [\lambda > \tfrac{1}{2}] \tag{38.01}$$

form a set of zero measure.

We shall prove this theorem by a reductio ad absurdum. Let there exist such a t, and let its binary expansion be

$$\frac{a_1}{2} + \frac{a_2}{4} + \cdots + \frac{a_n}{2^n} + \cdots \qquad (a_1, a_2, \cdots = 0 \text{ or } 1)\,. \tag{38.02}$$

Then if (38.01) holds good, we must have for every n

$$\begin{cases} \left| \psi(t, \alpha) - \psi\left(\frac{a_1}{2} + \frac{a_2}{4} + \cdots + \frac{a_n}{2^n}, \alpha\right) \right| 2^{\lambda n} < A < \infty, \\ \left| \psi(t, \alpha) - \psi\left(\frac{a_1}{2} + \frac{a_2}{4} + \cdots + \frac{a_n - 1}{2^n}, \alpha\right) \right| 2^{\lambda n} < A, \end{cases} \tag{38.03}$$

which gives us

$$\left| \psi\left(\frac{a_1}{2} + \cdots + \frac{a_n}{2^n}, \alpha\right) - \psi\left(\frac{a_1}{2} + \cdots + \frac{a_n - 1}{2^n}, \alpha\right) \right| 2^{\lambda n} < 2A\,. \tag{38.04}$$

If we can prove that the probability of such a contingency as (38.04) is zero, we shall have established our theorem.

To this end, we wish to discuss the distribution of $\psi(t + \epsilon, \alpha)$ when $\psi(t, \alpha)$ and $\psi(t + 2\epsilon, \alpha)$ are given. The measure of the set of values of α for which $\Re(\psi(t + \epsilon, \alpha) - \psi(t, \alpha))$ lies between u and $u + du$, while

$$\Re(\psi(t + 2\epsilon, \alpha) - \psi(t, \alpha))$$

lies between v and $v + dv$, is asymptotically

$$\frac{1}{2\pi^2 \epsilon} \exp\left(\frac{-u^2}{2\pi\epsilon} - \frac{(v - u)^2}{2\pi\epsilon}\right) \tag{38.05}$$

* *Notes on random functions*, R. E. A. C. Paley, N. Wiener, and A. Zygmund, Mathematische Zeitschrift, vol. 37 (1933), pp. 647–668. Cf. especially Theorem VII, p. 666. This theorem differs from Theorem XLVI of this book in that it concerns itself with $\chi(\alpha, x)$ which corresponds to the real part of $\psi(x, \alpha)$.

as du and $dv \to 0$. This can be seen by introducing appropriate f_1, f_2, and Φ in (37.30). Thus the ratio of the measure of the set of values of α for which $|u| < u_0$, and v lies between v_0 and $v_0 + dv$, to the measure of all values of α for which v lies between v_0 and $v_0 + dv$, u being taken to be $\Re(\psi(t + \epsilon, \alpha) - \psi(t, \alpha))$ and v to be $\Re(\psi(t + 2\epsilon, \alpha) - \psi(t, \alpha))$, is

$$\frac{\int_{-u_0}^{u} \exp\left(\frac{-u^2}{2\pi\epsilon} - \frac{(v_0 - u)^2}{2\pi\epsilon}\right) du}{\int_{-\infty}^{\infty} \exp\left(\frac{-u^2}{2\pi\epsilon} - \frac{(v_0 - u)^2}{2\pi\epsilon}\right) du} = \frac{\int_{-u_0}^{u_0} \exp\left(\frac{-2u^2 + 2uv_0}{2\pi\epsilon}\right) du}{\int_{-\infty}^{\infty} \exp\left(\frac{-2u^2 + 2uv_0}{2\pi\epsilon}\right) du}$$

$$\text{(38.06)} \qquad = \frac{\int_{-u_0 - v_0/2}^{u_0 - v_0/2} \exp\left(\frac{-2u^2}{2\pi\epsilon}\right) du}{\int_{-\infty}^{\infty} \exp\left(\frac{-2u^2}{2\pi\epsilon}\right) du}$$

$$\leqq 2\pi u_0 \epsilon^{1/2}.$$

An exactly similar result holds for the imaginary part of ψ. Thus if $\psi(t, \alpha)$ and $\psi(t + 2\epsilon, \alpha)$ are subject to any restrictions whatever, which hold good for a set of values of α of measure M, and if we further demand that

$$\text{(38.061)} \qquad |\psi(t + \epsilon, \alpha) - \psi(t, \alpha)| \leqq u_0 ,$$

it will follow that independently

$$\text{(38.062)} \qquad |\Re(\psi(t + \epsilon, \alpha) - \psi(t, \alpha))| \leqq u_0, \qquad |\Im(\psi(t + \epsilon, \alpha) - \psi(t, \alpha))| \leqq u_0 ,$$

and the measure of the set of values of α for which ψ is subject to the original restrictions together with (38.061) will not exceed $2u_0^2 \epsilon M$. The same result will apply if we replace (38.061) by

$$\text{(38.063)} \qquad |\psi(t + 2\epsilon, \alpha) - \psi(t + \epsilon, \alpha)| \leqq u_0 .$$

In each case we shall describe the situation by saying that whatever $\psi(t, \alpha)$ and $\psi(t + 2\epsilon, \alpha)$ may be, the *probability* of (38.062) or (38.063) does not exceed const. $u_0^2 \epsilon$. Thus the probability that of the 2^{q+1} quantities

$$|\psi((\mu + 1)2^{-q-1}, \alpha) - \psi(\mu 2^{-q-1}, \alpha)| ,$$

at least N_q do not exceed $2^{-\lambda_q} \cdot a$, whatever the distribution of the 2^q analogous quantities of the previous series, does not exceed

$$\text{(38.07)} \qquad \sum_{\nu = N}^{2^{q+1}} \frac{(2^{q+1})!}{\nu!\,(2^{q+1} - \nu)!} (1 - \text{const. } 2^{q-2\lambda_q}) (\text{const. } 2^{q-2\lambda_q})^{\nu} .$$

This quantity has the asymptotic representation

$$\text{(38.071)} \qquad \frac{1}{\pi^{1/2}} \int_{A_q}^{\infty} e^{-u^2}\, du ,$$

for large q, where

(38.072) $$A_q = \frac{N_q - \text{const. } 2^{2q+1-2\lambda q}}{\text{const. } 2^{q(1-\lambda)}}.$$

This results from familiar theorems concerning the relation between the Gaussian distribution and the binomial distribution.* Thus if

(38.08) $$N_q = \text{const. } 2^{q(2-2\mu)+1} \qquad [0 \leqq \mu < \lambda],$$

the expression (38.072) becomes greater than

(38.09) $$\text{const. } 2^{q(1-2\mu+\lambda)} - \text{const. } 2^{q(1-\lambda)},$$

which is greater than const. $2^{q(1-2\mu+\lambda)}$ and in turn greater than $2^{q(1-\mu)}$ for sufficiently large values of q. Now, if $\mu < 1$,

(38.10) $$\sum_1^\infty \frac{1}{\pi^{1/2}} \int_{\text{const.} 2^{q(1-\lambda)}}^\infty e^{-u^2}\, du < \infty.$$

Thus it is infinitely improbable that more than a finite number of q's can be found for which at least N_q of the 2^{q+1} quantities

(38.11) $$| \psi((\mu + 1)\, 2^{-q-1}, \alpha) - \psi(\mu 2^{-q-1}, \alpha) |$$

do not exceed const. $2^{-\lambda q}$. On the other hand, the probability that the S_q's consisting of all points in the N_q intervals of the qth stage have any point in common for all values of q from $Q_1 + 1$ to Q_2, does not exceed

(38.12) $$2^{Q_2+1} \prod_{q=Q_1+1}^{Q_2} \text{const. } \frac{2^{q(2-2\mu)}}{2^{q+1}} \leqq C_1 C_2^{Q_2}\, 2^{(1-2\mu)Q_2^2/2},$$

where C_1 and C_2 are constants. If $\mu > \frac{1}{2}$ this tends to 0 as $Q_2 \to \infty$. Thus it is infinitely improbable that there exists any point common to all the S_q's from some stage on, and (38.04) cannot hold true for any t and all n's. This establishes Theorem XLVI.

A counterpart to Theorem XLVI is

THEOREM XLVII.† *If $\lambda < \frac{1}{2}$, then except for a set of values of α of zero measure,*

(38.13) $$\lim_{\epsilon \to 0} (\psi(t + \epsilon, \alpha) - \psi(t, \alpha))/\epsilon^\lambda = 0$$

uniformly for all values of t.

* C. Jordan, *Statistique Mathématique*, Paris, 1927, pp. 105 ff.

† A particular case of this theorem has been proved by N. Wiener, in *Generalized harmonic analysis*, Acta Mathematica, vol. 55, pp. 219, 220.

Since the function $\psi(t, \alpha)$ is continuous in t, there is no essential limitation to (38.13) if we restrict it to terminating binary values of t, and let ϵ tend to zero through terminating binary values. Now let

$$2^{\lambda n} \mid \psi((k+1)2^{-n}, \alpha) - \psi(k2^{-n}, \alpha) \mid < A \tag{38.14}$$

for all n from some value on, and for all k in $(0, 2^n - 1)$.

Then if in the binary scale

$$t = \overbrace{.000 \cdots}^{(n-1)\text{ times}} a_n a_{n+1} \cdots \quad \text{and} \quad t + \epsilon = \overbrace{.000 \cdots}^{(n-1)\text{ times}} b_n b_{n+1} \cdots, \tag{38.15}$$

we can put

$$s = \begin{cases} .000 \cdots a_n c_{n+1} c_{n+2} \cdots; \\ .000 \cdots b_n d_{n+1} d_{n+2} \cdots \qquad (a_n = c_n; \qquad b_n = d_n). \end{cases} \tag{38.16}$$

Let us put

$$\begin{cases} S_m = .000 \cdots a_n a_{n+1} \cdots a_m, \\ T_m = .000 \cdots b_n b_{n+1} \cdots b_m. \end{cases} \tag{38.161}$$

Then

$$\begin{aligned} \mid \psi(t+\epsilon, \alpha) - \psi(t, \alpha) \mid &\leqq \sum_{k=n}^{\infty} \Big\{ \mid \psi(s_{k+1}, \alpha) - \psi(s_k, \alpha) \mid \\ &\quad + \mid \psi(T_{k+1}, \alpha) - \psi(T_k, \alpha) \mid \Big\} + \mid \psi(S_n, \alpha) - \psi(s, \alpha) \mid \\ &\quad + \mid \psi(T_n, \alpha) - \psi(s, \alpha) \mid \\ &\leqq 2A \sum_n^{\infty} 2^{-\lambda k} \\ &= \frac{2A\,2^{-\lambda n}}{1 - 2^{-\lambda}} \leqq \text{const.} \mid \epsilon^{\lambda} \mid, \end{aligned} \tag{38.17}$$

and (38.13) is established.

The probability that

$$2^{\lambda n} \mid \psi((k+1)\,2^{-n}, \alpha) - \psi(k\,2^{-n}, \alpha) \mid \geqq A \tag{38.18}$$

for a particular λ, a particular n, and a particular k, is by (37.48)

$$\int_{(A^2/(4\pi)^2)\,2^{(1-2\lambda)n}}^{\infty} e^{-u}\, du = e^{(-A^2/(4\pi^2))\,2^{(1-2\lambda)n}}. \tag{38.19}$$

If we sum this for $0 \leqq k < 2^n$ we get a result not exceeding

$$\exp\left(n \log 2 - \frac{A^2}{4\pi^2}\, 2^{(1-2\lambda)n}\right); \tag{38.20}$$

and if we now sum for all $n > N$, we get a result not exceeding

$$\sum_{N}^{\infty} \exp\left(n \log 2 - \frac{A^2}{4\pi^2} 2^{(1-2\lambda)n}\right); \tag{38.21}$$

and this tends to zero as $N \to \infty$. Thus except for a set of values of α of zero measure, there exist an $A > 0$ and an N such that if $n > N$ and $0 \leqq k < 2^n$, (38.16) is satisfied. We have seen this to lead to (38.15), uniformly for all values of t.

Theorems XLVI and XLVII lead, as it were, in opposite directions. The first holds when $\lambda > \frac{1}{2}$, the second when $\lambda < \frac{1}{2}$. In the critical case $\lambda = \frac{1}{2}$, more powerful tools are necessary for more than the simplest results. A simple result, proved in the same manner as Theorem XLVII, is that we may there substitute

$$\epsilon^{1/2}\left(-\frac{\log \epsilon}{\log 2}\right)^{1+\theta}$$

for ϵ^{λ}, if $\theta > 0$. In this case, important results have been obtained by Kolmogoroff.*

* A. Kolmogoroff, *Grundbegriffe der Warscheinlichkeitsrechnung*, Berlin, 1933.

Chapter X

The harmonic analysis of random functions

39. The ergodic theorem. In the discussion of random functions, one of the most important theorems is due to Birkhoff.* This theorem, which is in reality a general theorem from the theory of Lebesgue measure, lies at the basis of the study of ergodic dynamical systems. The simplest proof is that of Khintchine,† and depends on a method outlined by Hopf.‡

Let V be a portion of space of finite volume, which is subject to a stationary flow into itself. In what follows, x always denotes a point of V, and integration with respect to x is a space integration over V. If a moving particle is at time zero in the point x, then $T_\lambda x$ denotes the point where it is to be found at time λ. If M is an arbitrary sub-set of V, then $T_\lambda M$ will have the analogous significance. If the set M is Lebesgue measurable with measure $\mathfrak{M}(M)$ we assume that

$$\mathfrak{M}(T_\lambda M) = \mathfrak{M}(M) \tag{39.01}$$

for all λ.

If $f(x)$ is real and belongs to L over V, we put

$$\phi(x, t) = \int_0^t f(T_\lambda x)\, d\lambda\,. \tag{39.02}$$

Clearly, this line integral has a sense for almost all x. Birkhoff's theorem now states that *as t becomes infinite,*

$$\lim_{t \to \infty} \frac{1}{t}\, \phi(x, t) \tag{39.03}$$

exists as a function of x for almost all x.

40. The theory of transformations. This chapter concerns the study of certain means associated with the Brownian motion, more especially where these means involve a harmonic analysis in the complex domain. Our theory is thus an exact counterpart to the theory of almost periodic functions in the complex domain, as developed by Bohr and Jessen.§ As a necessary tool, we

* G. D. Birkhoff, Proceedings of the National Academy of Sciences, vol. 17 (1931), pp. 650–660.

† A. Khintchine, *Zu Birkhoff's Lösung des Ergodenproblems*, Mathematische Annalen, vol. 107, pp. 485–488.

‡ E. Hopf, *Complete transitivity and the ergodic principle*, Proceedings of the National Academy of Sciences, vol. 18 (1932), pp. 204–209.

§ H. Bohr und B. Jessen, *Über die Werteverteilung der Riemannschen Zetafunktion*, Erste Mitteilung, Acta Mathematica, vol. 54 (1930), pp. 31–35, Zweite Mitteilung, vol. 58 (1932), pp. 51–55.

employ the theory of unitary transformations in Hilbert space. There is a strong resemblance between the order of ideas here introduced and the theory of mixture of Eberhard Hopf,* and indeed we might make our entire theory a special case of his. After careful consideration, we have decided against this mode of presentation, which involves the substitution for the form of Hilbert space most directly associated with our problem, of another somewhat factitious one. We wish to thank Professors Hopf and Jessen more particularly for having pointed out the great simplification in our argument which may be made by an introduction of Birkhoff's fundamental theorem.

We have already seen that $\psi(x, \alpha)$ has the formal trigonometric series

$$\psi(x, \alpha) \sim x(-\log \alpha_1)^{1/2} e^{2\pi i \alpha_2} + \sum_1^\infty \frac{e^{inx}}{in} (-\log \alpha_{4n-1})^{1/2} e^{2\pi i \alpha_{4n}} + \sum_1^\infty \frac{e^{-inx}}{-in} (-\log \alpha_{4n+1})^{1/2} e^{2\pi i \alpha_{4n+2}}. \tag{37.061}$$

That is,

$$\begin{aligned} \frac{1}{2\pi} \int_{-\pi}^{\pi} d\psi(x, \alpha) &= (-\log \alpha_1)^{1/2} e^{2\pi i \alpha_2}; \\ \frac{1}{2\pi} \int_{-\pi}^{\pi} e^{-inx} d\psi(x, \alpha) &= (-\log \alpha_{4n-1})^{1/2} e^{2\pi i \alpha_{4n}} \qquad (n = 1, 2, \cdots), \\ \frac{1}{2\pi} \int_{-\pi}^{\pi} e^{inx} d\psi(x, \alpha) &= (-\log \alpha_{4n+1})^{1/2} e^{2\pi i \alpha_{4n+2}} \qquad (n = 1, 2, \cdots). \end{aligned} \tag{40.01}$$

Since the determination of the sequence $\{\alpha_k\}$ is equivalent to the determination of α, there is a one-one correspondence between α and the sequence

$$\frac{1}{2\pi} \int_{-\pi}^{\pi} e^{inx} d\psi(x, \alpha) \qquad [n = \cdots, -2, -1, 0, 1, 2, \cdots]. \tag{40.015}$$

Now let T be a unitary transformation of Hilbert space into itself. Then if $\{\phi_n(x)\}$ is a normal and orthogonal closed sequence, so is $\{T\phi_n(x)\}$ and $\{T^{-1}\phi_n(x)\}$. Let $\phi_n(x) = e^{inx}/(2\pi)^{1/2}$; then the sequences $Te^{inx}/(2\pi)^{1/2}$ and $T^{-1}e^{inx}/(2\pi)^{1/2}$ are normal, orthogonal, and closed. It follows from Theorem XLIV that if α is given, then (except at the most in a null set of cases) the sequences

$$\int_{-\pi}^{\pi} Te^{inx} d\psi(x, \alpha) \quad \text{and} \quad \int_{-\pi}^{\pi} T^{-1} e^{inx} d\psi(x, \alpha)$$

are determined. Let us define β by

$$\int_{-\pi}^{\pi} e^{inx} d\psi(x, \beta) = \int_{-\pi}^{\pi} Te^{inx} d\psi(x, \alpha). \tag{40.02}$$

* E. Hopf, *Complete transitivity and the ergodic principle*, Proceedings of the National Academy of Sciences, vol. 18 (1932), pp. 204–209.

This determines β as a unique function of α for almost all values of α. We have already seen that any measurable function of an infinite set of variables may be approximated L_1 as closely as we wish by a function of a finite number of these variables. Any measurable function of α is a function of the variables $\int_{-\pi}^{\pi} Te^{inx}\, d\psi(x, \alpha)$; and if we put

$$(40.03) \qquad \frac{1}{2\pi}\int_{-\pi}^{\pi} Te^{inx}\, d\psi(x, \alpha) = (-\log \beta_j)^{1/2}\, e^{2\pi i \beta_k}\,,$$

where if $n = 1, j = 1, k = 2$; if $n > 0, j = 4n + 1, k = 4n + 2$; if $n = -m < 0, j = 4m - 1, k = 4m$, the β_ν's will be independent of one another and will be distributed uniformly over the interval (0, 1) (cf. (37.27)). It will readily follow that the integral of any function of β depending only on a finite number of β_ν's will be the same as the integral of that function with respect to α, and hence, by the approximation theorem just mentioned, the integral of any measurable function of β with respect to β will be the same as its integral with respect to α.

Again, it follows from (40.02) that

$$(40.04) \qquad \int_{-\pi}^{\pi} \sum_{-N}^{N} a_n\, e^{inx}\, d\psi(x, \beta) = \int_{-\pi}^{\pi} T\left(\sum_{-N}^{N} a_n\, e^{inx}\right) d\psi(x, \alpha)\,,$$

from which an argument such as that of Theorem XLIV′ will prove that if $F(x)$ belongs to L_2,

$$(40.05) \qquad \int_{-\pi}^{\pi} F(x)\, d\psi(x, \beta) = \int_{-\pi}^{\pi} T\, F(x)\, d\psi(x, \alpha)\,.$$

In particular,

$$(40.06) \qquad \int_{-\pi}^{\pi} T^{-1}\, e^{inx}\, d\psi(x, \beta) = \int_{-\pi}^{\pi} e^{inx}\, d\psi(x, \alpha)\,.$$

This enables us to determine α as a unique function of β for almost all values of β. Thus T determines a transformation of α into β which is measure-preserving and which is one-one for almost all values of α and for almost all values of β. We shall write

$$(40.07) \qquad \psi(x, \alpha) = T\psi(x, \beta)\,.$$

Such a transformation allows us to apply the Birkhoff theorem. The null sets of values of α and of β for which the mapping ceases to be one-one may be replaced by other null sets which map in a one-one manner, and which have no influence over Birkhoff's theorem.

We now introduce the group of unitary transformations T^t determined by the property $T^tT^u = T^{t+u}$. Applying Birkhoff's theorem, if

$$\Phi\left\{\int_{-\pi}^{\pi} T^tF(x)\, d\psi(x, \alpha)\right\}$$

is Lebesgue integrable in α, then for almost all α,

$$\lim_{A\to\infty} \frac{1}{A} \int_0^A \Phi\left\{\int_{-\pi}^{\pi} T^t F(x)\, d\psi(x,\alpha)\right\} dt \tag{40.08}$$

exists. This will be the case whenever (37.26) converges absolutely.

Now let us consider

$$\begin{aligned}
&\int_0^1 \left|\frac{1}{A}\int_0^A \Phi\left\{\int_{-\pi}^{\pi} T^t F(x)\, d\psi(x,\alpha)\right\} dt - \int_0^1 \Phi\left\{\int_{-\pi}^{\pi} F(y)\, d\psi(y,\beta)\right\} d\beta\right|^2 d\alpha \\
&= \int_0^1 \left|\frac{1}{A}\int_0^A \Phi\left\{\int_{-\pi}^{\pi} T^t F(x)\, d\psi(x,\alpha)\right\} dt\right|^2 d\alpha - \left|\int_0^1 \Phi\left\{\int_{-\pi}^{\pi} F(y)\, d\psi(y,\beta)\right\} d\beta\right|^2 \\
&= \frac{1}{A^2}\int_0^A ds \int_0^A dt \int_0^1 \Phi\left\{\int_{-\pi}^{\pi} T^s F(x)\, d\psi(x,\alpha)\right\} \bar{\Phi}\left\{\int_{-\pi}^{\pi} T^t F(y)\, d\psi(y,\alpha)\right\} d\alpha \\
&\qquad - \left|\int_0^1 \Phi\left\{\int_{-\pi}^{\pi} F(y)\, d\psi(y,\beta)\right\} d\beta\right|^2 .
\end{aligned} \tag{40.09}$$

Let us apply Theorem XLIV′. If we put

$$\int_{-\pi}^{\pi} |F(x)|^2\, dx = U, \qquad \int_{-\pi}^{\pi} \overline{F(x)}\, T^{s-t} F(x)\, dx = V, \tag{40.091}$$

then the right side of (40.09) becomes

$$\begin{aligned}
&\frac{1}{A^2}\int_0^A ds \int_0^A dt \int_{-\infty}^{\infty} du_1 \int_{-\infty}^{\infty} dv_1 \int_{-\infty}^{\infty} du_2 \int_{-\infty}^{\infty} dv_2\, \Phi(u_1+iv_1)\, \bar{\Phi}(u_2+iv_2) \\
&\times\left[\frac{\exp\left[\dfrac{-(u_1^2+u_2^2+v_1^2+v_2^2)\,U + 2(u_1u_2+v_1v_2)\,\Re(V) + 2(u_1v_2-u_2v_1)\,\Im(V)}{U^2-|V|^2}\right]}{4\pi^4[U^2-|V|^2]}\right. \\
&\qquad \left. - \frac{\exp\left[\dfrac{-(u_1^2+u_2^2+v_1^2+v_2^2)}{U}\right]}{4\pi^4 U^2}\right].
\end{aligned} \tag{40.092}$$

Let us suppose that Φ is subject to conditions which make

$$\begin{aligned}
&\int_{-\infty}^{\infty} du_1 \int_{-\infty}^{\infty} dv_1 \int_{-\infty}^{\infty} du_2 \int_{-\infty}^{\infty} dv_2\, \Phi(u_1+iv_1)\, \bar{\Phi}(u_2+iv_2) \\
&\times\left[\frac{\exp\left[\dfrac{-(u_1^2+u_2^2+v_1^2+v_2^2)\,U + 2(u_1u_2+v_1v_2)\,\Re(V) + 2(u_1v_2-v_1u_2)\,\Im(V)}{U^2-|V|^2}\right]}{4\pi^4[U^2-|V|^2]}\right.
\end{aligned} \tag{40.093}$$

$$- \frac{\exp\left[-\dfrac{(u_1^2+u_2^2+v_1^2+v_2^2)}{U}\right]}{4\pi^4 U^2}\Bigg]$$

bounded if $U < P$, $|V| < Q < P$. Then if

$$\lim_{u\to\infty}\int_{-\pi}^{\pi}\overline{F(x)}\,T^u F(x)\,dx = 0\,, \tag{40.094}$$

we shall have

$$\begin{aligned}
&\lim_{A\to\infty}\int_0^1\Bigg|\frac{1}{A}\int_0^A \Phi\left\{\int_{-\pi}^{\pi} T^t F(x)\,d\psi(x,\alpha)\right\}dt \\
&\qquad - \frac{1}{2\pi}\int_{-\pi}^{\pi} d\theta\int_0^\infty e^{-u}\,\Phi\left\{e^{i\theta}\left[2\pi u\int_{-\pi}^{\pi}|F(x)|^2\,dx\right]^{1/2}\right\}du\Bigg|^2 d\alpha \\
&= \lim_{B\to 0}\int_{-\infty}^{\infty}du_1\int_{-\infty}^{\infty}du_2\int_{-\infty}^{\infty}dv_1\int_{-\infty}^{\infty}dv_2\,\Phi(u_1+iv_1)\,\bar\Phi(u_2+iv_2) \\
&\quad\times\left[\frac{\exp\left[\dfrac{-(u_1^2+u_2^2+v_1^2+v_2^2)+2(u_1u_2+v_1v_2)\,\Re(B)+2(u_1v_2-u_2v_1)\,\Im(B)}{U(1-|B|^2)}\right]}{4\pi^4U^2(1-|B|^2)}\right. \\
&\qquad\left. - \frac{\exp\left[\dfrac{-(u_1^2+u_2^2+v_1^2+v_2^2)}{U}\right]}{4\pi^4U^2}\right] = 0\,.
\end{aligned} \tag{40.095}$$

Thus by (40.08) and the fact that a limit and a limit in the mean of the same function differ at most over a null set of arguments, we have for almost all α,

$$\begin{aligned}
&\lim_{A\to\infty}\frac{1}{A}\int_0^A \Phi\left\{\int_{-\pi}^{\pi} T^t F(x)\,d\psi(x,\alpha)\right\}dt \\
&\qquad = \frac{1}{2\pi}\int_{-\pi}^{\pi} d\theta\int_0^\infty e^{-u}\,\Phi\left\{e^{i\theta}\left[2\pi u\int_{-\pi}^{\pi}|F(x)|^2\,dx\right]^{1/2}\right\}du\,.
\end{aligned} \tag{40.096}$$

For example, if (40.05) holds, we have for almost all α,

$$\begin{aligned}
&\lim_{A\to\infty}\frac{1}{2A}\int_{-A}^{A}\left|\int_{-\pi}^{\pi} T^t F(x)\,d\psi(x,\alpha)\right|^w dt \\
&\qquad = \Gamma\left(\frac{w}{2}+1\right)\left\{2\pi\int_{-\pi}^{\pi}|F(x)|^2\,dx\right\}^{w/2}.
\end{aligned} \tag{40.097}$$

This result is a special case of the following one, which may be proved by the same method:

THEOREM XLVIII. *Let $F_1(x), \cdots, F_N(x)$ be a set of functions of L_2, and let T^t be a linear group of unitary transformations with the property that for all m and n,*

$$\lim_{t\to\infty} \int_{-\pi}^{\pi} \overline{F_m(x)}\, T^t F_n(x)\, dx = 0. \tag{40.098}$$

Then if

$$\Phi\left\{ \int_{-\pi}^{\pi} F_1(x)\, d\psi(x, \alpha), \cdots, \int_{-\pi}^{\pi} F_N(x)\, d\psi(x, \alpha) \right\} \tag{40.10}$$

is Lebesgue integrable with respect to α,

$$\begin{aligned} &\lim_{A\to\infty} \frac{1}{A} \int_0^A \Phi\left\{ \int_{-\pi}^{\pi} T^t F_1(x)\, d\psi(x, \alpha), \cdots, \int_{-\pi}^{\pi} T^t F_N(x)\, d\psi(x, \alpha) \right\} dt \\ &\quad = \int_0^1 \Phi\left\{ \int_{-\pi}^{\pi} F_1(x)\, d\psi(x, \beta), \cdots, \int_{-\pi}^{\pi} F_N(x)\, d\psi(x, \beta) \right\} d\beta \end{aligned} \tag{40.11}$$

for almost all α.

A particularly important transformation group T^t is the group of all translations on the infinite line. A group of transformations on $(-\pi, \pi)$ isomorphic with this under the transformation $y = \tan(x/2)$ is defined as follows: let us put

$$F(x) = 2^{-1/2} f(\tan(x/2)) \sec(x/2); \quad G(x) = 2^{-1/2} g(\tan(x/2)) \sec(x/2); \tag{40 12}$$

and

$$T^t F(x) = 2^{-1/2} f(\tan(x/2) + t) \sec(x/2). \tag{40.13}$$

The group in question is manifestly unitary. If now $F(x)$ and $G(x)$ are *any* two functions belonging to L_2, we shall have

$$\lim_{t\to\infty} \int_{-\pi}^{\pi} \overline{F(x)}\, T^t G(x)\, dx = \lim_{t\to\infty} \int_{-\infty}^{\infty} \overline{f(\xi)}\, g(\xi + t)\, d\xi = 0. \tag{40.14}$$

Let us here introduce the notation

$$\int_{-\pi}^{\pi} F(x)\, d\psi(x, \alpha) = \int_{-\infty}^{\infty} f(\xi)\, d\Psi(\xi, \alpha). \tag{40.15}$$

We shall formally have

$$\begin{aligned} \int_{-\pi}^{\pi} F(x)\, d\psi(x, \alpha) &= \int_{-\pi}^{\pi} 2^{-1/2} f(\tan(x/2)) \sec(x/2)\, d\psi(x, \alpha) \\ &= \int_{-\infty}^{\infty} 2^{-1/2} f(\xi)\, (1 + \xi^2)^{1/2}\, d\psi(2 \tan^{-1} \xi, \alpha). \end{aligned} \tag{40.16}$$

If we put

$$(40.17)\quad \begin{aligned}\Psi(\xi, \alpha) &= \int_0^{\xi} 2^{-1/2}(1+\eta^2)^{1/2}\, d\psi(2\tan^{-1}\eta, \alpha)\\ &= \int_0^{\tan(\xi/2)} 2^{-1/2}\sec(x/2)\, d\psi(x, \alpha)\end{aligned}$$

in accordance with this notation, we shall be consistent. We shall have

$$(40.18)\quad \begin{aligned}\int_{-\pi}^{\pi} T^t F(x)\, d\psi(x, \alpha) &= \int_{-\infty}^{\infty} 2^{-1/2} f(\xi + t)(1+\xi^2)^{1/2}\, d\psi(2\tan^{-1}\xi, \alpha)\\ &= \int_{-\infty}^{\infty} f(\xi+t)\, d\Psi(\xi, \alpha),\end{aligned}$$

which we may reasonably write

$$(40.19)\quad \int_{-\infty}^{\infty} f(\xi)\, d\Psi(\xi - t, \alpha).$$

By Theorem XLVIII, if

$$\Phi\left\{\int_{-\pi}^{\pi} T^t e^{in_1x}\, d\psi(x, \alpha), \cdots, \int_{-\pi}^{\pi} T^t e^{in_kx}\, d\psi(x, \alpha)\right\} dt$$

is any functional of $\psi(x, \alpha)$ which depends only on a finite number of Fourier coefficients of $\psi(x, \alpha)$ and is integrable with regard to α, we shall have for almost all α

$$(40.20)\quad \begin{aligned}&\lim_{A\to\infty} \frac{1}{A}\int_a^{A+a} \Phi\left\{\int_{-\pi}^{\pi} T^t e^{in_1x}\, d\psi(x, \alpha), \cdots, \int_{-\pi}^{\pi} T^t e^{in_kx}\, d\psi(x, \alpha)\right\} dt\\ &= \lim_{T\to\infty} \frac{1}{A}\int_a^{A+a} \Phi\left\{\int_{-\pi}^{\pi} e^{in_1x}\, dT^{-t}\psi(x, \alpha), \cdots, \int_{-\pi}^{\pi} e^{in_kx}\, dT^{-t}\psi(x, \alpha)\right\},\end{aligned}$$

which is equal to a fixed value independent of α, for almost all α. Now, we have seen that every functional of $\psi(x, \alpha)$ which is a continuous function of α is the uniform limit of a sequence of integrable functionals depending only on a finite set of Fourier coefficients of $\psi(x, \alpha)$, and any integrable function of α is except for a set of values of α of arbitrarily small measure the uniform limit of a sequence of continuous functions. From this we obtain at once

THEOREM XLIX. *Let T^t be defined as in* (40.12) *and* (40.13). *Let $\mathfrak{F}(\psi(x, \alpha))$ be any functional of $\psi(x, \alpha)$ which is integrable with respect to α. Then for almost all α,*

$$(40.21)\quad \frac{1}{N}\int_a^{N+a} \mathfrak{F}(\psi(x+t, \alpha))\, dt$$

tends as $N \to \infty$ to the limit $\int_0^1 \mathfrak{F}(\psi(x, \alpha))\, d\alpha$.

41. **The harmonic analysis of random functions.** Let $f(x+iy)$ be an analytic function such that $(1+x^2)f(x+iy)$ and $(1+|x|)f'(x+iy)$ belong uniformly to L_2 over $a < y < b$. For any enumerable set of y's and almost all α's, we shall have as a consequence of Theorem XLIX,

$$\lim_{N\to\infty} \frac{1}{2N} \int_{-N}^{N} dx \left| \int_{-\infty}^{\infty} f(\xi + x + iy)\, d\Psi(\xi, \alpha) \right|^2$$

$$= \int_0^1 d\alpha \left| \int_{-\infty}^{\infty} f(\xi+x+iy)\, d\Psi(\xi, \alpha) \right|^2$$

$$= \int_0^1 d\alpha \left| \int_{-\pi}^{\pi} T^t[2^{-1/2} f(\tan(u/2) + x + iy) \sec(u/2)]\, d\psi(u, \alpha) \right|^2$$

$$(41.01) \quad = \frac{1}{2\pi} \int_{-\pi}^{\pi} d\theta \int_0^{\infty} e^{-u} \left| e^{i\theta} \times \left[2\pi u \int_{-\pi}^{\pi} | 2^{-1/2} f(\tan(w/2) + x + iy) \sec(w/2) |^2\, dw \right]^{1/2} \right|^2 du$$

$$= \int_0^{\infty} 2\pi u \int_{-\infty}^{\infty} | f(\xi + iy) |^2\, d\xi\, e^{-u}\, du$$

$$= 2\pi \int_{-\infty}^{\infty} | f(\xi + iy) |^2\, d\xi\, .$$

In exactly the same way, if $a \leqq y_0 < y_1 \leqq b$, we shall have

$$(41.02) \quad \lim_{N\to\infty} \frac{1}{2N} \int_{-N}^{N} dx \int_{y_0}^{y_1} dy \left| \int_{-\infty}^{\infty} f(\xi + x + iy)\, d\Psi(\xi, \alpha) \right|^2 = 2\pi \int_{y_0}^{y_1} dy \int_{-\infty}^{\infty} | f(\xi + iy) |^2\, d\xi\, .$$

Furthermore,

$$\int_{-\infty}^{\infty} f(\xi + x + iy)\, d\Psi(\xi, \alpha)$$

$$(41.03) \quad = \int_{-\infty}^{\infty} 2^{-1/2} f(\xi + x + iy)\, (1 + \xi^2)^{1/2}\, d\psi(2 \tan^{-1} \xi, \alpha)$$

$$= \int_{-\pi}^{\pi} 2^{-1/2} f(\tan(u/2) + x + iy) \sec(u/2)\, d\psi(u, \alpha)$$

$$= -\int_{-\pi}^{\pi} \psi(u, \alpha) \frac{d}{du} (2^{-1/2} f(\tan(u/2) + x + iy) \sec(u/2))$$

converges absolutely and uniformly for almost all α since ψ is bounded and continuous, and thus (41.03) is analytic over $a < y < b$.

By Cauchy's theorem and the Schwarz inequality

$$
\begin{aligned}
(41.04)\quad & \left| \int_{-\infty}^{\infty} f(\xi + x_0 + iy_0)\, d\Psi(\xi, \alpha) \right| = \frac{1}{2\pi} \left| \int_C \frac{\int_{-\infty}^{\infty} f(\xi + x + iy)\, d\Psi(\xi, \alpha)}{x - x_0 + i(y - y_0)}\, ds \right| \\
& \leqq \frac{r}{2\pi} \left[\int_0^{2\pi} \frac{d\theta}{|x - x_0|^2 + |y - y_0|^2} \int^{2\pi} \left| \int_{-\infty}^{\infty} f(\xi + x + iy)\, d\Psi(\xi, \alpha) \right|^2 d\theta \right]^{1/2},
\end{aligned}
$$

where r is the radius of the circle C, and θ is the angle subtended at the center of this circle. A comparison with (41.01) will show that uniformly in x over $a + \epsilon < y < b - \epsilon$,

$$
(41.05)\quad \lim_{N\to\infty} \frac{1}{2N} \int_{-N}^{N} dx \left| \int_{-\infty}^{\infty} f(x + \xi + iy)\, d\Psi(\xi, \alpha) \right|^2 < \text{const.}
$$

Furthermore, along any finite or denumerable set of abscissae, for almost all α, by Theorem XLVIII,

$$
\begin{aligned}
(41.06)\quad & \lim_{T\to\infty} \frac{1}{2T} \int_{-T}^{T} dx \left[\int_{-\infty}^{\infty} f(\xi + x + iy + t)\, d\Psi(\xi, \alpha) \int_{-\infty}^{\infty} \overline{f(\xi + x + iy)\, d\Psi(\xi, \alpha)} \right] \\
& = 2\pi \int_{-\infty}^{\infty} f(\xi + iy + t)\, \bar{f}(\xi + iy)\, d\xi\,.
\end{aligned}
$$

Thus by the results of chapter VIII,

$$
\int_{-\infty}^{\infty} f(x + iy)\, d\Psi(x, \alpha)
$$

belongs to S' over $a < y < b$ for almost all α and is bounded over any interval $y_0 \leqq y \leqq y_1$. Thus it follows that (41.01) holds for every y in (y_0, y_1) for almost every α.

Let us suppose that

$$
(41.07)\quad f(z) = \underset{A\to\infty}{\text{l.i.m.}} \frac{1}{2\pi} \int_{-A}^{A} \phi(u)\, e^{iuz}\, du
$$

along each abscissa for $a < y < b$. It will then follow from (34.40) of chapter VIII that for almost all α, all t, and all y in (y_0, y_1),

$$
\begin{aligned}
(41.08)\quad & \lim_{T\to\infty} \frac{1}{2T} \int_{-T}^{T} dx \left[\int_{-\infty}^{\infty} f(\xi + x + iy + t)\, d\Psi(\xi, \alpha) \int_{-\infty}^{\infty} \overline{f(\xi + x + iy)\, d\Psi(\xi, \alpha)} \right] \\
& = \frac{1}{2\pi} \int_{-\infty}^{\infty} |\,\phi(u)\,|^2\, e^{iut - uy}\, du\,.
\end{aligned}
$$

42. The zeros of a random function. Let $f(x + iy)$ have the properties attributed to it in the last section. Then we shall have by Theorem XLIX

$$\lim_{A\to\infty}\frac{1}{2A}\int_{-A}^{A}dx\int_{y_0}^{y_1}dy\left|\log\left|\int_{-\infty}^{\infty}f(\xi+x+iy)\,d\Psi(\xi,\alpha)\right|\right| \tag{42.01}$$
$$=\int_0^\infty e^{-u}\,du\int_{y_0}^{y_1}dy\left|\log\left|\left[2\pi u\int_{-\infty}^{\infty}|f(x+iy)|^2\,dx\right]^{1/2}\right|\right|.$$

Accordingly, for any C

$$\lim_{A\to\infty}\frac{1}{2A}\left[\int_{-A-C}^{-A}+\int_{A}^{A+C}\right]dx\int_{y_0}^{y_1}dy\left|\log\left|\int_{-\infty}^{\infty}f(\xi+x+iy)\,d\Psi(\xi,\alpha)\right|\right|=0. \tag{42.02}$$

It also follows from the properties of a harmonic function that

$$\lim_{A\to\infty}\frac{1}{2A}\int_{-A}^{A}dx\int_{y_0}^{y_1}dy\log\left|\int_{-\infty}^{\infty}f(\xi+x+iy)\,d\Psi(\xi,\alpha)\right|$$
$$=\lim_{A\to\infty}\frac{1}{2A}\int_{-A}^{A}dx\,2\int_0^{(y_1-y_0)/2}\frac{r\,dr}{\left[\left(\frac{y_1-y_0}{2}\right)^2-r^2\right]^{1/2}} \tag{42.03}$$
$$\times\frac{1}{2\pi}\int_0^{2\pi}\log\left|\int_{-\infty}^{\infty}f\left(\xi+x+\frac{i(y_0+y_1)}{2}+r\,e^{i\theta}\right)d\Psi(\xi,\alpha)\right|d\theta.$$

On the other hand, let $\phi(z)$ be bounded over $|\Im(z)|\leqq a$ and measurable in $\Re(z)$ and $\Im(z)$. By the transformation of polar into Cartesian coordinates,

$$\frac{1}{2A}\int_{-A}^{A}dx\,2\int_0^a[a^2-r^2]^{-1/2}\,r\,dr\,\frac{1}{2\pi}\int_0^{2\pi}\phi(x+r\,e^{i\theta})\,d\theta \tag{42.04}$$
$$=\frac{1}{2A}\int_{-A}^{A}dx\,\frac{1}{\pi}\int_{-a}^{a}d\eta\int_{-[a^2-\eta^2]^{1/2}}^{[a^2-\eta^2]^{1/2}}\phi(x+\xi+i\eta)\,[a^2-\xi^2-\eta^2]^{-1/2}\,d\xi.$$

The change of variables $w = x + \xi$ will reduce this to

$$\frac{1}{2A}\int_{-a}^{a}d\eta\,\frac{1}{\pi}\int_{-[a^2-\eta^2]^{1/2}}^{[a^2-\eta^2]^{1/2}}d\xi\int_{-A+\xi}^{A+\xi}\phi(w+i\eta)\,[a^2-\xi^2-\eta^2]^{-1/2}\,dw$$
$$=\frac{1}{2A}\int_{-a}^{a}d\eta\left\{\int_{-A}^{A}\phi(w+i\eta)\,dw\,\frac{1}{\pi}\int_{-[a^2-\eta^2]^{1/2}}^{[a^2-\eta^2]^{1/2}}[a^2-\xi^2-\eta^2]^{-1/2}\,d\xi\right.$$
$$+\frac{1}{\pi}\int_{-[a^2-\eta^2]^{1/2}}^{[a^2-\eta^2]^{1/2}}d\xi\int_{A}^{A+\xi}\phi(w+i\eta)\,[a^2-\xi^2-\eta^2]^{-1/2}\,dw \tag{42.041}$$
$$\left.+\frac{1}{\pi}\int_{-[a^2-\eta^2]^{1/2}}^{[a^2-\eta^2]^{1/2}}d\xi\int_{-A-\xi}^{-A}\phi(w+i\eta)\,[a^2-\xi^2-\eta^2]^{-1/2}\,dw\right\}$$
$$=\frac{1}{2A}\int_{-a}^{a}d\eta\int_{-A}^{A}\phi(w+i\eta)\,dw+R,$$

where

(42.042)
$$|R| \leqq \text{const.}\ \frac{1}{2A}\int_{-a}^{a} d\eta \int_{-[a^2-\eta^2]^{1/2}}^{[a^2-\eta^2]^{1/2}} d\xi\, |\xi|\, [a^2 - \xi^2 - \eta^2]^{-1/2} \leqq \text{const.}\ A^{-1}.$$

Thus

(42.043)
$$\frac{1}{2A}\int_{-A}^{A} dx\, 2\int_0^a [a^2 - r^2]^{-1/2}\, r\, dr\, \frac{1}{2\pi}\int_0^{2\pi} \phi(x + r\,e^{i\theta})\, d\theta = \frac{1}{2A}\int_{-A}^{A} dw \int_{-a}^{a} \phi(w + i\eta)\, d\eta + O(A^{-1}).$$

If we combine this with (42.02), we get

(42.05)
$$\lim_{A\to\infty}\frac{1}{2A}\int_{-A}^{A} dx\, 2\int_0^{(y_1-y_0)/2} \frac{r\,dr}{\left[\left(\frac{y_1-y_0}{2}\right)^2 - r^2\right]^{1/2}} \times\left[\frac{1}{2\pi}\int_0^{2\pi}\log\left|\int_{-\infty}^{\infty} f\left(\xi + x + \frac{i(y_0+y_1)}{2} + r\,e^{i\theta}\right) d\Psi(\xi,\alpha)\right| d\xi - \log\left|\int_{-\infty}^{\infty} f\left(\xi + x + \frac{i(y_0+y_1)}{2}\right) d\Psi(\xi,\alpha)\right|\right]$$
$$= \int_0^{\infty} e^{-u}\,du\int_{y_0}^{y_1} dy\left\{\log\left|\left[2\pi u\int_{-\infty}^{\infty}|f(x+iy)|^2\,dx\right]^{1/2}\right| \cdot - \log\left|\left[2\pi u\int_{-\infty}^{\infty}\left|f\left(x + \frac{i(y_0+y_1)}{2}\right)\right|^2 dx\right]^{1/2}\right|\right\}$$
$$= \frac{1}{2}\int_{y_0}^{y_1} dy\,\log\left\{\frac{\int_{-\infty}^{\infty}|f(x+iy)|^2\,dx}{\int_{-\infty}^{\infty}\left|f\left(x+\frac{i(y_0+y_1)}{2}\right)\right|^2 dx}\right\}.$$

On the other hand, by Jensen's theorem, if $\xi_n + i\eta_n$ are the zeros of $\int_{-\infty}^{\infty} f(\xi + z)\, d\Psi(\xi, \alpha)$ in the strip $y_0 \leqq \Im(z) \leqq y_1$,

(42.06)
$$\frac{1}{2\pi}\int_0^{2\pi}\log\left|\int_{-\infty}^{\infty} f\left(\xi + x + \frac{i(y_0+y_1)}{2} + r\,e^{i\theta}\right) d\Psi(\xi,\alpha)\right| d\theta - \log\left|\int_{-\infty}^{\infty} f\left(\xi + x + \frac{i(y_0+y_1)}{2}\right) d\Psi(\xi,\alpha)\right|$$
$$= \sum_{J(r)} \log\left\{r\left[(\xi_n - x)^2 + \left(\eta_n - \frac{y_0+y_1}{2}\right)^2\right]^{-1/2}\right\},$$

where $J(r)$ is the region characterized by

(42.061)
$$(\xi_n - x)^2 + \left(\eta_n - \frac{y_0+y_1}{2}\right)^2 \leqq r^2.$$

Thus

(42.07)
$$\frac{1}{2}\int_{y_0}^{y_1} dy \log\left\{\frac{\int_{-\infty}^{\infty} | f(x+iy) |^2 dx}{\int_{-\infty}^{\infty}\left| f\left(x+\frac{i(y_0+y_1)}{2}\right)\right|^2 dx}\right\}$$
$$= \lim_{A\to\infty}\frac{1}{2A}\int_{-A}^{A} dx \int_0^{(y_1-y_0)/2}\frac{r\,dr}{[((y_1-y_0)/2)^2 - r^2]^{1/2}}$$
$$\times \sum_J \log\left\{r\left[(\xi_n - x)^2 + \left(\eta_n - \frac{y_0+y_1}{2}\right)^2\right]^{-1/2}\right\}.$$

Let us put

(42.071)
$$\left[(\xi_n - x)^2 + \left(\eta_n - \frac{y_0+y_1}{2}\right)^2\right]^{1/2} = \rho_n^2 .$$

Then the left side of (42.07) becomes

(42.072)
$$\lim_{A\to\infty}\frac{1}{A}\int_{-A}^{A} dx \sum_{J((y_1-y_0)/2)}\int_{\rho_n}^{(y_1-y_0)/2} r\left[\left(\frac{y_1-y_2}{2}\right)^2 - r^2\right]^{-1/2}\log\frac{r}{\rho_n}\,dr .$$

Now let us make use of the formula

(42.08)
$$\int_b^a \log\frac{r}{b}\frac{r\,dr}{(a^2-r^2)^{1/2}} = \int_b^a (a^2-r^2)^{1/2}\frac{dr}{r}$$
$$= -a\left[\log\left(\frac{a}{b} - \left(\frac{a^2}{b^2}-1\right)^{1/2}\right) + \left(1-\frac{b^2}{a^2}\right)^{1/2}\right].$$

If we substitute this in (42.07–42.072), we get

(42.09)
$$\frac{1}{2}\int_{y_0}^{y_1} dy \log\left\{\frac{\int_{-\infty}^{\infty} | f(x+iy) |^2 dx}{\int_{-\infty}^{\infty}\left| f\left(x+\frac{i(y_0+y_1)}{2}\right)\right|^2 dx}\right\}$$
$$= \lim_{A\to\infty}\frac{1}{A}\int_{-A}^{A} dx \sum_{J((y_1-y_0)/2)}\frac{y_0-y_1}{2}\left\{\log\left[\frac{y_1-y_0}{2\rho_n} - \left(\frac{(y_1-y_0)^2}{4\rho_n^2}-1\right)^{1/2}\right]\right.$$
$$\left. + \left[1-\frac{4\rho_n^2}{(y_1-y_0)^2}\right]^{1/2}\right\}.$$

Let us now invert the order of summation and integration. The boundaries of our combined integration and summation mean that x lies between $-A$ and A, and that (ξ_n, η_n) lies within a circle of radius $\frac{1}{2}(y_1 - y_0)$ about $(x, \frac{1}{2}(y_0 + y_1))$. This implies that ξ_n lies between limits depending on η_n, but intermediate between $(-A + \frac{1}{2}(y_1 - y_0), A - \frac{1}{2}(y_1 - y_0))$ and $(-A - \frac{1}{2}(y_1 - y_0), A + \frac{1}{2}(y_1 - y_0))$. Furthermore, the integrand of (42.09) is always positive.

Since the limit in (42.09) exists, we shall get the same limit if we replace A by $A \pm \frac{1}{2}(y_1 - y_0)$, which bears a ratio to A tending to 1 as $A \to \infty$. Accordingly, if we take the range of ξ_n to be from $-A$ to A, we shall not modify the limit in (42.09). The point $(x, \frac{1}{2}(y_0 + y_1))$ then lies within a circle K of radius $\frac{1}{2}(y_1 - y_0)$ about (ξ_n, η_n). The left side of (42.09) becomes accordingly

$$\lim_{A\to\infty} \frac{1}{A} \sum_{|\xi_n|<A} \int_K \frac{y_0 - y_1}{2} \left\{ \log \left[\frac{y_1 - y_0}{2\rho_n} - \left(\frac{(y_1 - y_0)^2}{4\rho_n^2} - 1 \right)^{1/2} \right] + \left[1 - \frac{4\rho_n^2}{(y_1 - y_0)^2} \right]^{1/2} \right\} dx. \tag{42.091}$$

Let us now put $u = \xi_n - x$, and carry out an integration by parts. The right side of (42.09) becomes

$$\lim_{A\to\infty} \frac{1}{A} \sum_{|\xi_n|<A} \int_L \frac{u^2 \left[\left(\frac{y_1 - y_0}{2} \right)^2 - u^2 - \left(\eta_n - \frac{y_0 + y_1}{2} \right)^2 \right]^{1/2}}{\frac{y_1 - y_0}{2} \left(u^2 + \left(\eta_n - \frac{y_0 + y_1}{2} \right)^2 \right)} du, \tag{42.092}$$

where L is the interior of the circle

$$u^2 + \left(\eta_n - \frac{y_0 + y_1}{2} \right)^2 \leqq \left(\frac{y_1 - y_0}{2} \right)^2.$$

If we now put

$$w = \left[u^2 + \left(\eta_n - \frac{y_0 + y_1}{2} \right)^2 \right]^{1/2}; \tag{42.093}$$

this gives us for the right side of (42.09)

$$\lim_{A\to\infty} \frac{2}{A} \sum_{|\xi_n|<A} \int_{|\eta_n - (y_0+y_1)/2|}^{(y_1 - y_0)/2} \left\{ \left[w^2 - \left(\eta_n - \frac{y_0 + y_1}{2} \right)^2 \right] \times \left[\left(\frac{y_1 - y_0}{2} \right)^2 - w^2 \right] \right\}^{1/2} \frac{dw}{w}. \tag{42.094}$$

We now make use of the formula

$$\int_q^p [(x^2 - q^2)(p^2 - x^2)]^{1/2} \frac{dx}{x} = \frac{\pi}{4} (p - q)^2. \tag{42.10}$$

Then the right side of (42.09) becomes

$$\begin{aligned} &\lim_{A\to\infty} \frac{\pi}{2A} \sum_{|\xi_n|<A} \left[\frac{y_1 - y_0}{2} - \left| \eta_n - \frac{y_0 + y_1}{2} \right| \right]^2 \\ &= \pi \lim_{A\to\infty} \frac{1}{2A} \sum_{|\xi_n|<A} \text{smaller of } (\eta_n - y_0)^2, (\eta_n - y_1)^2, \end{aligned} \tag{42.11}$$

which always exists if y_0 and y_1 are interior to the interval $a < y < b$ over any interior interval of which $f(x + iy)$ uniformly belongs to L_2 and satisfies the conditions following (41.06) in the last section. Over $a + \epsilon < y < b - \epsilon$ we have uniformly

$$\overline{\lim_{A\to\infty}} \frac{1}{2A} \sum_{|\xi_n|<A} 1 < \text{const.}\ (y_1 - y_0)\,, \tag{42.12}$$

in accordance with (42.09) and (42.11), since $(\eta_n - y_0)^2 > \epsilon^2$, $(\eta_n - y_1)^2 > \epsilon^2$. A further result of these formulas is

$$\begin{aligned} &\frac{1}{2\pi\epsilon}\left[\int_{y_1-\epsilon}^{y_1} + \int_{y_0}^{y_0+\epsilon}\right] \log \left\{ \frac{\int_{-\infty}^{\infty} |f(x+iy)|^2\,dx}{\int_{-\infty}^{\infty} \left| f\left(x + i\frac{y_0+y_1}{2}\right)\right|^2 dx} \right\} \\ &\quad = \lim_{A\to\infty} \frac{1}{2A}\Bigg\{ - \sum_{|\xi_n^{\text{i}}|<A} (2\eta_n^{\text{i}} - 2y_1 + \epsilon) + \sum_{|\xi_n^{\text{ii}}|<A} (2\eta_n^{\text{ii}} + 2y_0 - \epsilon) \\ &\qquad + \frac{1}{\epsilon}\sum_{|\xi^{\text{iii}}|<A} (\eta_n^{\text{iii}} - y_0)^2 + \frac{1}{\epsilon}\sum_{|\xi_n^{\text{iv}}|<A} (\eta_n^{\text{iv}} - y_1)^2 \Bigg\}. \end{aligned} \tag{42.13}$$

Here $\{\xi_n^{\text{i}} + i\eta_n^{\text{i}}\}$ are the zeros of $\int_{-\infty}^{\infty} f(\xi + z)\,d\Psi(\xi; \alpha)$ in

$$\frac{y_0 + y_1}{2} \leqq \mathfrak{J}(z) \leqq y_1 - \epsilon\,.$$

$\{\xi_n^{\text{ii}} + i\eta_n^{\text{ii}}\}$ are the zeros in $(y_0 + \epsilon, \frac{1}{2}(y_0 + y_1))$; $\{\xi_n^{\text{iii}} + i\eta_n^{\text{iii}}\}$ are the zeros in $(y_0, y_0 + \epsilon)$, and $\{\xi_n^{\text{iv}} + i\eta_n^{\text{iv}}\}$ are the zeros in $(y_1 - \epsilon, y_1)$.

Let us remember that

$$\int_{-\infty}^{\infty} |f(x + iy)|^2\,dx = \frac{1}{2\pi}\int_{-\infty}^{\infty} |\phi(u)|^2 e^{-2uy}\,du\,, \tag{42.14}$$

which is continuous in y, and that

$$\frac{d}{dy}\int_{-\infty}^{\infty} |f(x + iy)|^2\,dx = -\frac{1}{\pi}\int_{-\infty}^{\infty} u\,|\phi(u)|^2 e^{-2uy}\,du\,. \tag{42.15}$$

It follows then from (42.13), if we let $\epsilon \to 0$ and take (42.12) into consideration, that

$$\begin{aligned} &\frac{1}{4\pi}\log \frac{\int_{-\infty}^{\infty} |f(x + iy_0)|^2\,dx \int_{-\infty}^{\infty} |f(x + iy_1)|^2\,dx}{\left[\int_{-\infty}^{\infty} \left| f\left(x + \frac{i(y_0+y_1)}{2}\right)\right|^2 dx\right]^2} \\ &\quad = \lim_{A\to\infty}\frac{1}{2A}\Bigg\{ - \sum_{|\xi_n^{\text{i}}|<A} (\eta_n^{\text{i}} - y_1) + \sum_{|\xi_n^{\text{ii}}|<A} (\eta_n^{\text{ii}} - y_0) \Bigg\}. \end{aligned} \tag{42.16}$$

Now let us replace y_0 by $y_0 + \epsilon$ and y_1 by $y_1 - \epsilon$. We have

$$\frac{1}{4\pi} \log \frac{\int_{-\infty}^{\infty} |f(x + i(y_0 + \epsilon))|^2 \, dx \int_{-\infty}^{\infty} |f(x + i(y_1 - \epsilon))|^2 \, dx}{\left[\int_{-\infty}^{\infty} \left| f\left(x + \frac{i(y_0 + y_1)}{2}\right)\right|^2 dx\right]^2}$$

$$= \lim_{A\to\infty} \frac{1}{2A}\left\{- \sum_{|\xi_n^{\text{i}}|<A} (\eta_n^{\text{i}} - y_1 + \epsilon) + \sum_{|\xi_n^{\text{ii}}|<A} (\eta_n^{\text{ii}} - y_0 - \epsilon)\right\}, \tag{42.161}$$

where $\{\xi_n^{\text{i}} + i\eta_n^{\text{i}}\}$ represents the zeros of $\int_{-\infty}^{\infty} f(\xi + z) \, d\Psi(\xi, \alpha)$ in $\frac{1}{2}(y_0 + y_1) \leqq \Im(z) \leqq y_1 - \epsilon$ and $\{\xi_n^{\text{ii}} + i\eta_n^{\text{ii}}\}$ represents the zeros in $(y_0 + \epsilon, \frac{1}{2}(y_0 + y_1))$. If we interpret $\{\xi_n^{\text{iii}} + i\eta_n^{\text{iii}}\}$ and $\{\xi_n^{\text{iv}} + i\eta_n^{\text{iv}}\}$ as above, we see that

$$\frac{1}{4\pi} \log \frac{\int_{-\infty}^{\infty} |f(x + iy_0)|^2 \, dx \int_{-\infty}^{\infty} |f(x + iy_1)|^2 \, dx}{\int_{-\infty}^{\infty} |f(x + i(y_0 + \epsilon))|^2 \, dx \int_{-\infty}^{\infty} |f(x + i(y_1 - \epsilon))|^2 \, dx}$$

$$= \lim_{A\to\infty} \left\{\frac{-\epsilon}{2A}\left\{\sum_{|\xi_n^{\text{i}}|<A} 1 + \sum_{|\xi_n^{\text{ii}}|<A} 1\right\}\right. \tag{42.162}$$

$$\left. + \frac{1}{2A}\left\{\sum_{|\xi_n^{\text{iii}}|<A} (\eta_n^{\text{iii}} - y_1 + \epsilon) - \sum_{|\xi_n^{\text{iv}}|<A} (\eta_n^{\text{iv}} - y_0 - \epsilon)\right\}\right\}.$$

As before, by the use of (42.12),

$$\lim_{A\to\infty} \frac{1}{2A} \sum_{|\xi_n|<A} 1$$

$$= \lim_{\epsilon\to 0} \frac{1}{4\pi\epsilon} \log \frac{\int_{-\infty}^{\infty} |f(x + i(y_0 + \epsilon))|^2 \, dx \int_{-\infty}^{\infty} |f(x + i(y_1 - \epsilon))|^2 \, dx}{\int_{-\infty}^{\infty} |f(x + iy_0)|^2 \, dx \int_{-\infty}^{\infty} |f(x + iy_1)|^2 \, dx} \tag{42.163}$$

$$= \frac{1}{4\pi}\left\{\left[\frac{d}{dy} \log \int_{-\infty}^{\infty} |f(x + iy)|^2 \, dx\right]_{y=y_0} - \left[\frac{d}{dy} \log \int_{-\infty}^{\infty} |f(x + iy)|^2 \, dx\right]_{y=y_1}\right\}.$$

Thus by (42.15) and (42.14),

$$\frac{1}{4\pi}\left\{\frac{\int_{-\infty}^{\infty} - 2u \, |\phi(u)|^2 e^{-2uy_1} \, du}{\int_{-\infty}^{\infty} |\phi(u)|^2 e^{-2uy_1} \, du} - \frac{\int_{-\infty}^{\infty} - 2u \, |\phi(u)|^2 e^{-2uy_0} \, du}{\int_{-\infty}^{\infty} |\phi(u)|^2 e^{-2uy_0} \, du}\right\} \tag{42.17}$$

$$= \lim_{A\to\infty} \frac{1}{2A} \sum_{|\xi_n|<A} 1.$$

We may then state our result as

THEOREM L. *Let*

$$\int_{-\infty}^{\infty} (1 + x^2)^2 \, | f(x + iy) |^2 \, dx < \text{const.} \qquad (a < y < b), \tag{42.18}$$

and let

$$\int_{-\infty}^{\infty} (1 + x^2) \, | f'(x + iy) |^2 \, dx < \text{const.} \qquad (a < y < b). \tag{42.19}$$

Let

$$f(z) = \underset{A \to \infty}{\text{l.i.m.}} \frac{1}{2\pi} \int_{-A}^{A} \phi(u) \, e^{iuz} \, du \tag{42.20}$$

along each abscissa in $a < y < b$. Then the number of zeros of

$$\int_{-\infty}^{\infty} f(\xi + z) \, d\Psi(\xi, \alpha) \tag{42.21}$$

in the strip $y_0 < \Im(z) < y_1$ ($a < y_0 < y_1 < b$) between the abscissas $-A$ and A is asymptotically

$$\frac{A}{2\pi} \left\{ \frac{\int_{-\infty}^{\infty} 2u \, | \phi(u) |^2 \, e^{-2uy_0} \, du}{\int_{-\infty}^{\infty} | \phi(u) |^2 \, e^{-2uy_0} \, du} - \frac{\int_{-\infty}^{\infty} 2u \, | \phi(u) |^2 \, e^{-2uy_1} \, du}{\int_{-\infty}^{\infty} | \phi(u) |^2 \, e^{-2uy_1} \, du} \right\}. \tag{42.22}$$

This theorem forms the exact counterpart for random functions of certain recent theorems of Bohr and Jessen for almost periodic functions.* The two classes of theorems depend on a use of Jensen's formula. Jessen uses a special Jensen formula for the zeros on an infinite strip, which involves means, and may be proved by arguments approximating to such a region by the region between two circles on the conformal map of such a region. We use the classical Jensen formula, integrating it with respect to parameters determining the center of the radius of the circle to which it applies. It is believed that our methods can be applied directly to the type of problem studied by Bohr and Jessen.

* Cf. note by B. Jessen, to appear in the Journal of Mathematics and Physics of the Massachusetts Institute of Technology.

BIBLIOGRAPHY

A. S. Besicovitch, *Almost Periodic Functions.* Cambridge, 1932, p. 10.

G. D. Birkhoff, Proceedings of the National Academy of Sciences (U. S. A.), vol. 17 (1931), pp. 650–660.

A theorem on series of orthogonal functions with an application to Sturm-Liouville series. Proceedings of the National Academy of Sciences (U. S. A.), vol. 3 (1917), p. 656.

S. Bochner, *Inversion formulae and unitary transformations.* Annals of Mathematics, (2), vol. 34 (1934), pp. 111–115.

Vorlesungen über Fouriersche Integrale. Leipzig, 1932.

Integration von Funktionen, deren Werte die Elemente eines Vektorraumes sind. Fundamenta Mathematicae, vol. 20 (1933), pp. 262–276.

H. Bohr and B. Jessen, *Über die Werteverteilung der Riemannschen Zetafunktion.* Acta Mathematica, vol. 54 (1930), pp. 1–35; vol. 58 (1932), pp. 1–55.

E. Borel, *Les probabilités dénombrables et leurs applications arithmétiques.* Rendiconti del Circolo Matematico di Palermo, vol. 27 (1909), pp. 247–271.

I. W. Busbridge, *On general transforms of the Fourier type.* Journal of the London Mathematical Society, vol. 9 (1934), pp. 179–187.

T. Carleman, *Les Fonctions Quasi-Analytiques.* Paris, 1926.

P. J. Daniell, *A general form of integral.* Annals of Mathematics, (2), vol. 19 (1918), pp. 279–294.

Integrals in an infinite number of dimensions. Annals of Mathematics, (2), vol. 20 (1919), pp. 281–288.

Further properties of the general integral. Annals of Mathematics, (2), vol. 21 (1920), pp. 203–220.

M. Denjoy, Comptes Rendus, vol. 173 (1921), p. 1329.

P. Dienes, *The Taylor Series.* Oxford, 1931, pp. 372 ff.

A. Einstein, Annalen der Physik, vol. 17 (1905), pp. 549 ff.; vol. 19 (1906), pp. 371 ff.

G. H. Hardy, *A theorem concerning Fourier transforms.* Journal of the London Mathematical Society, vol. 8 (1933), pp. 227–231.

G. H. Hardy and E. C. Titchmarsh, *A class of Fourier kernels.* Proceedings of the London Mathematical Society, (2), vol. 35 (1932), pp. 116–155.

E. Hopf, *Über lineare Integralgleichungen mit positivem Kern.* Sitzungsberichte der Berliner Akademie der Wissenschaften, 1928, No. XVIII.

Mathematisches zur Strahlungsgleichgewichtstheorie der Fixsternatmosphären. Mathematische Zeitschrift, vol. 33 (1931), p. 109.

Complete transitivity and the ergodic principle. Proceedings of the National Academy of Sciences (U. S. A.), vol. 18 (1932), pp. 204–209.

A. E. Ingham, *The Distribution of Primes.* Cambridge Tracts in Mathematics and Mathematical Physics, No. 30, Ch. III, §7.

B. Jessen, Paper to appear in the Journal of Mathematics and Physics of the Massachusetts Institute of Technology.

Bidrag til Integralteorien for Funktioner af unendelig mange Variable. Copenhagen, 1930.

C. Jordan, *Statistique Mathématique.* Paris, 1927, pp. 105 ff.

A. Khintchine, *Zu Birkhoff's Lösung des Ergodenproblems.* Mathematische Annalen, vol. 107 (1933), pp. 285–288.

K. Knopp, *Theory and Application of Infinite Series.* London, 1928.

A. Kolmogoroff, *Grundbegriffe der Warscheinlichkeitsrechnung.* Berlin, 1933.

E. Landau, *Beiträge zur analytischen Zahlentheorie.* Rendiconti del Circolo Matematico di Palermo, vol. 26 (1908), p. 218.

Darstellung und Begründung einiger neuerer Ergebnisse der Funktionentheorie. 1929.

N. Levinson, *On a theorem of Carleman.* Proceedings of the National Academy of Sciences (U. S. A.), vol. 20 (1934), pp. 523–525.

P. Lévy, *Sur la convergence absolue des séries de Fourier.* Comptes Rendus, vol. 196 (1933), p. 463.

M. S. Mandelbrojt, *Sur l'unicite des series de Fourier.* Journal de l'Ecole Polytechnique, (2), cahier 32 (1934).

J. Mercer, *On the limits of real variants.* Proceedings of the London Mathematical Society, (2), vol. 5 (1907), pp. 206–224.

G. W. Morgan, *A note on Fourier transforms.* Journal of the London Mathematical Society, vol. 9 (1934), pp. 187–193.

C. H. Müntz, *Über den Approximationssatz von Weierstrass.* Schwarz's Festschrift, Berlin, 1914, pp. 303–312.

R. E. A. C. Paley and N. Wiener, *Notes on the theory and application of Fourier transforms.* Notes I–II, Transactions of the American Mathematical Society, vol. 35 (1933), pp. 348–355; Notes III–VII, ibid., vol. 35 (1933), pp. 761–791.

R. E. A. C. Paley, N. Wiener and A. Zygmund, *Notes on random functions.* Mathematische Zeitschrift, vol. 37 (1933), pp. 647–668.

R. E. A. C. Paley and A. Zygmund, *On some series of functions.* Proceedings of the Cambridge Philosophical Society, vol. 26, pp. 337–357, 458–474; vol. 28, pp. 190–205.

J. Perrin, *Atoms.* Translated by D. Ll. Hammick. Second English edition. London, 1923, pp. 109 ff.

M. Plancherel, *Sur les formules de réciprocité du type de Fourier.* Journal of the London Mathematical Society, vol. 8 (1933), pp. 220–226.

Contribution à l'étude de la répresentation d'une fonction arbitraire par des intégrales définies. Rendiconti del Circolo Matematico di Palermo, vol. 30 (1910), pp. 289–335.

G. Pólya, Jahresbericht der Deutschen Mathematiker-Vereinigung, vol. 40 (1931), 2te Abteiling, p. 80, Problem 105; also Problem 108 (p. 81).

H. Rademacher, *Einige Sätze über Reihen von allgemeinen Orthogonalfunktionen.* Mathematische Annalen, vol. 87 (1922), pp. 112–138.

L. L. Silverman, *On the consistency and equivalence of certain generalized definitions of the limit of a function of a continuous variable.* Annals of Mathematics, (2), vol. 21 (1920), pp. 128–140.

M. von Smoluchowski, Die Naturwissenschaften, vol. 6 (1918), pp. 253–263.

H. Steinhaus, *Sur la probabilité de la convergence de séries.* Studia Mathematica, vol. 2 (1930), pp. 21–39.

W. Stepanoff, *Sur quelques généralisations des fonctions presque-périodiques.* Comptes Rendus, vol. 181, pp. 90–92.

O. Szász, *Über die Approximation stetiger Funktionen durch gegebene Funktionenfolgen.* Mathematische Annalen, vol. 104 (1931), pp. 155–160.

Über die Approximation stetiger Funktionen durch lineare Aggregate von Potenzen Mathematische Annalen, vol. 77 (1916), pp. 482–496.

E. C. Titchmarsh, *The zeros of certain integral functions.* Proceedings of the London Mathematical Society, (2), vol. 25 (1926), pp. 283–302.

On integral functions with real negative zeros. Proceedings of the London Mathematical Society, (2), vol. 26 (1927), pp. 185–200.

A proof of a theorem of Watson. Journal of the London Mathematical Society, vol. 8 (1933), pp. 217–220.

The Theory of Functions. Oxford, 1932.

C. J. de la Vallée Poussin, Comptes Rendus, vol. 176 (1923), p. 635.

V. VOLTERRA, *Leçons sur les Equations Intégrales et les Equations Intégro-Différentielles.* Paris, 1913.

J. L. WALSH, *A generalization of the Fourier cosine series.* Transactions of the American Mathematical Society, vol. 22 (1921), pp. 230–239.

G. N. WATSON, *General transforms.* Proceedings of the London Mathematical Society, (2), vol. 35 (1932), pp. 156–199.

J. M. WHITTAKER, *On the cardinal function of interpolation theory.* Proceedings of the Edinburgh Mathematical Society, (2), vol. 1 (1927), pp. 41–47.

The "Fourier" theory of the cardinal function. Ibid., (2), vol. 1 (1928), pp. 169–177.

The lower order of integral functions. Journal of the London Mathematical Society, vol. 8 (1933), pp. 20–27.

D. V. WIDDER, *The inversion of the Laplace integral and the related moment problem.* Transactions of the American Mathematical Society, vol. 36 (1934), pp. 107–201.

N. WIENER, *A new method in Tauberian theorems.* Journal of Mathematics and Physics of the Massachusetts Institute of Technology, vol. 7 (1928), pp. 161–184.

Tauberian theorems. Annals of Mathematics, (2), vol. 33 (1932), pp. 1–100.

Generalized harmonic analysis. Acta Mathematica, vol. 55 (1930), pp. 117–258.

The Fourier Integral and Certain of its Applications. Cambridge, 1933.

On the closure of certain assemblages of trigonometrical functions. Proceedings of the National Academy of Sciences (U. S. A.), vol. 13 (1927), p. 27.

N. WIENER AND E. HOPF, *Über eine Klasse singulärer Integralgleichungen.* Sitzungsberichte der Berliner Akademie der Wissenschaften, 1931, p. 696.

INDEX